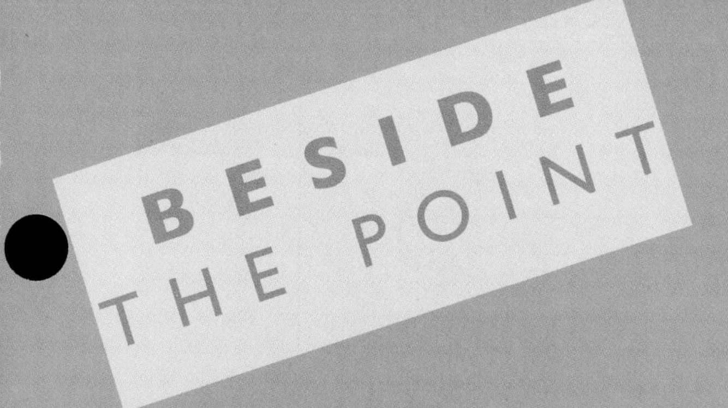

OPERATING WITH DECIMALS, PERCENTS, AND INTEGERS

0.1 $\frac{1}{10}$ 10%

MathScape™
SEEING AND THINKING MATHEMATICALLY

Unit Overview 206

Phase One Overview . 208
Lesson 1 210
Lesson 2 212
Lesson 3 214
Lesson 4 216

Phase Two Overview . 218
Lesson 5 220
Lesson 6 222
Lesson 7 224
Lesson 8 226
Lesson 9 228

Phase Three Overview 230
Lesson 10 232
Lesson 11 234
Lesson 12 236
Lesson 13 238
Lesson 14 240

Phase Four Overview . 242
Lesson 15 244
Lesson 16 246
Lesson 17 248
Lesson 18 250
Lesson 19 252
Lesson 20 254

Homework 256

PHASE**ONE**

Decimals

You probably have some change in your pocket right now. In this phase, you will investigate how money relates to our decimal system. You will be building on what you know about equivalent fractions to help you make "cents" of it all.

How can you use your own number sense to solve fraction, decimal, percent, and integer problems?

BESIDE THE POINT

PHASE**TWO**
Computing with Decimals

Decimals are all around us in the real world. In this phase, you will work on computation strategies for addition, subtraction, multiplication, and division of decimals. You will use what you know about the relationship between fractions and decimals to help you find efficient computation methods.

PHASE**THREE**
Percents

This phase explores the relationship among fractions, decimals, and percents. You will calculate percents to help interpret data. You will investigate percents that are less than 1% and percents that are greater than 100%. Finally, you will investigate how percents are commonly used and how they are sometimes misused.

PHASE**FOUR**
The Integers

You have probably heard negative numbers being used when discussing topics such as games and temperatures. In this phase, you will learn how negative numbers complete the integers. You will use number lines and cubes to learn how to add and subtract with negative numbers.

PHASE ONE

How is all your loose change related to decimals? How can you take what you know about money to help to make sense of decimals? This phase will help you make connections between fractions and decimals.

Decimals

WHAT'S THE MATH?

Investigations in this section focus on:

NUMBER AND COMPUTATION

- Expressing money in decimal notation
- Using area models to interpret and represent decimals
- Writing fractions as decimals and decimals as fractions
- Comparing fractions and decimals

ESTIMATION

- Rounding decimals

1 The Fraction-Decimal Connection

USING DECIMALS IN MONEY NOTATION

You are already familiar with decimals, because you often see them in daily life. Every time you use money or compare prices, you use decimals. To begin this in-depth study of decimals, you will relate familiar coins to decimal numbers.

How are fractions and decimals related?

Make Penny Rectangles

Using the Paper Pennies handout, work with a partner to determine how many different ways you can arrange 100 pennies in the shape of a rectangle. Each row must have the same number of pennies.

Make a table like the one below. Use it to organize your answers. Some answers may be the same rectangle, oriented a different way.

Describe the rectangle made with 100 pennies.	Write one row as a fraction of the whole.	Write the simplified fraction.	Write the value of one row of money.
4 rows of 25 pennies 4×25	$\frac{25}{100}$	$\frac{1}{4}$	$0.25

Use Area Models

How can decimals be represented using area models?

You can use area models to represent fractions and decimals.

1 Write the fraction represented by the shaded area of each square or squares. Then, write the corresponding decimal.

a. 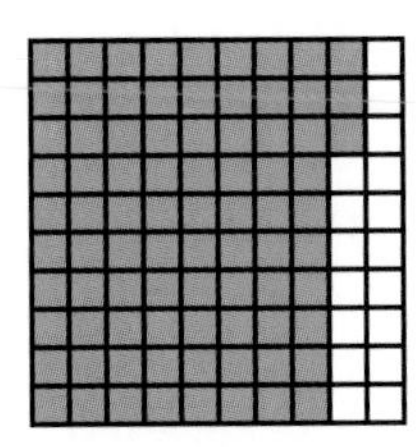

b. 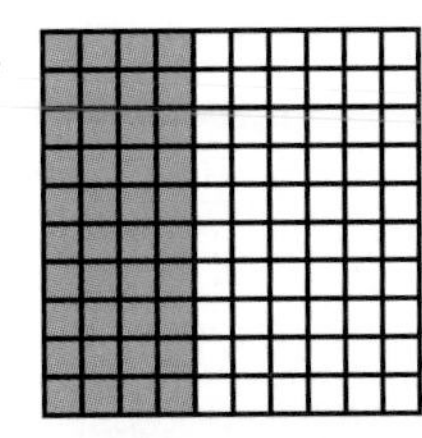

c. 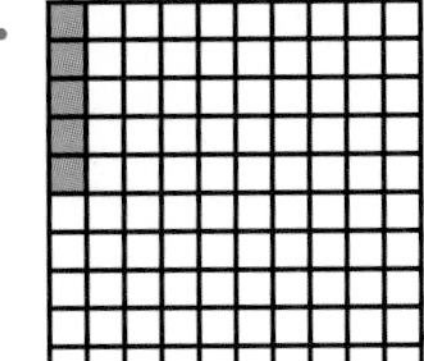

d. 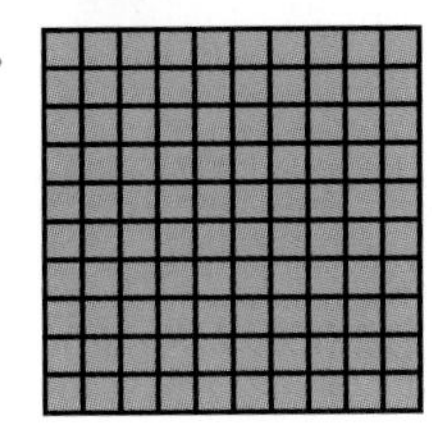 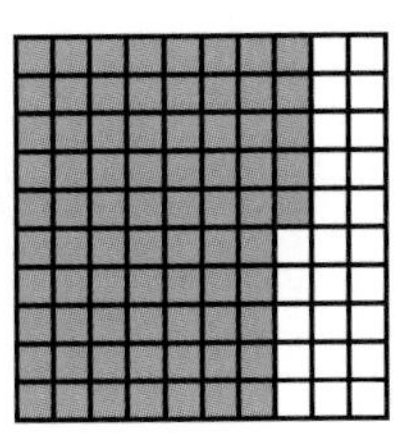

2 Draw an area model for each decimal.

a. 0.41 **b.** 0.95 **c.** 1.10 **d.** 1.38

Solve a Decimal Puzzle

Now, you and a partner will make and solve decimal puzzles. Cut out squares that are 10 centimeters by 10 centimeters.

- Divide your square into at least four parts. Shade each part a different color. Try to make each part a different shape and size.
- Carefully cut out the parts along the lines to create puzzle pieces. Trade puzzles with your partner.
- Describe each piece of your partner's puzzle as a fraction of the whole and as a decimal. Then, try to put the puzzle together.
- Write two number sentences that describe the puzzle. Use fractions in one number sentence and decimals in the other.

decimal system

page 256

2 What's the Point?

WRITING FRACTIONS AS DECIMALS

To write fractions as decimals in Lesson 1, you had to use place value. Now, you will write some more fractions as decimals. But, some of these decimals will require more decimal places.

Write Fractions as Decimals

How can you write a fraction as a decimal?

Use the place-value chart below to help you write decimals.

1 Write two different numbers that both have a 7 in the tens place, a 3 in the hundreds place, a 0 in the tenths place, a 4 in the hundredths place, and a 2 in the ones place. The two numbers will need to differ in other places.

2 Write each number as a decimal.

a. $13\frac{7}{10}$ **b.** $23\frac{4}{10}$ **c.** $2\frac{13}{100}$

3 Use what you know about equivalent fractions to write each number as a decimal.

a. $25\frac{1}{2}$ **b.** $12\frac{3}{5}$ **c.** $3\frac{8}{25}$

4 How could you write the number 17 so that it has a tenths digit, but still represents the same number?

Decimal Place Value

ten thousands 10,000	thousands 1,000	COMMA (for thousands period)	hundreds 100	tens 10	ones 1	DECIMAL POINT	tenths 0.1 $\frac{1}{10}$	hundredths 0.01 $\frac{1}{100}$	thousandths 0.001 $\frac{1}{1,000}$	ten thousandths 0.0001 $\frac{1}{10,000}$	hundred thousandths 0.00001 $\frac{1}{100,000}$
	4	,	7	2	1	.	0	3	8		

What number is shown in the diagram?

What digit is in the tenths place?

Find a Conversion Method

Study the patterns below.

Pattern I		Pattern II	
$4{,}000 \div 5 = 800$	$\frac{4{,}000}{5} = 800$	$1{,}000 \div 8 = 125$	$\frac{1{,}000}{8} = 125$
$400 \div 5 = 80$	$\frac{400}{5} = 80$	$100 \div 8 = 12.5$	$\frac{100}{8} = 12.5$
$40 \div 5 = 8$	$\frac{40}{5} = 8$	$10 \div 8 = 1.25$	$\frac{10}{8} = 1.25$
$4 \div 5 = ?$	$\frac{4}{5} = ?$	$1 \div 8 = ?$	$\frac{1}{8} = ?$

Can you devise a method for changing any fraction to a decimal?

1 How could you change $\frac{4}{5}$ and $\frac{1}{8}$ to a decimal without using equivalent fractions?

2 Use your method to write each fraction as a decimal. Use a calculator to do the calculations.

a. $\frac{5}{8}$ **b.** $3\frac{3}{8}$ **c.** $\frac{9}{16}$

d. $\frac{7}{32}$ **e.** $\frac{13}{40}$ **f.** $\frac{5}{64}$

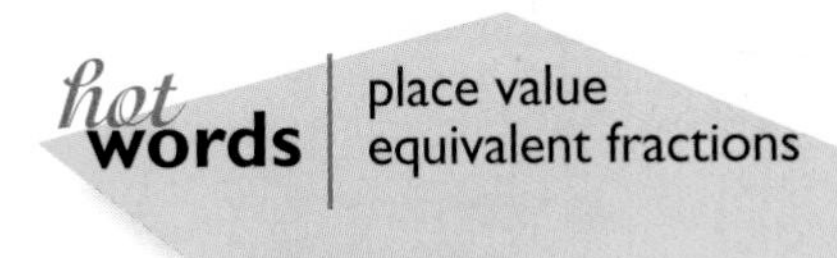

page 257

3 Put Them in Order

COMPARING AND ORDERING DECIMALS

You see decimals used every day, and you frequently need to compare them. In this lesson, you will first complete a handout to learn how to compare decimals using a number line. Then, you will learn to compare decimals without using a number line. Finally, you will play a game in which strategy helps you to create winning numbers.

Compare Decimals

How can you compare decimals?

Chris and Pat's teacher asked them to determine whether 23.7 or 3.17 is greater.

Pat said, "I lined them up and compared the digits. It is just like comparing 237 and 317. Since 3 is greater than 2, I know 3.17 is greater than 23.7."

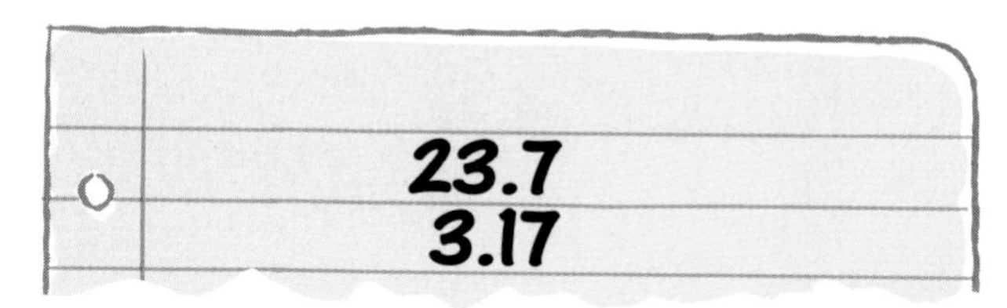

Chris listened to Pat's explanation and said, "I agree that 317 is greater than 237. But 3.17 is not greater than 23.7, because 3.17 is a little more than 3 and 23.7 is more than 23. I do not think you lined up the numbers correctly."

1. Do you agree with Pat or with Chris? How would you explain which of the two numbers is greater? How would you line up the numbers? Write a rule for comparing decimal numbers. Explain why your rule works.

2. Use your rule to place the following numbers in order from least to greatest. If two numbers are equal, place an equals sign between them.

   18.1 18.01 18.10 1.80494 2.785 0.2785 18.110

Play the "Place Value" Game

How can you create the least and greatest decimals?

The "Place Value" game is a game for two players.

1 Play a sample round of the "Place Value" game. How many points did each player get? Analyze the possibilities.

a. Suppose your partner's digit placements remained the same. Could you have placed your digits differently and earned two points?

b. Suppose your digit placements remained the same. Could your partner have placed his or her digits differently and earned two points?

c. Choosing from the six digits you drew, what is the greatest number you could make? What is the least number?

d. Choosing from the six digits your partner drew, what is the greatest number he or she could make? What is the least number?

2 Play the game until someone gets 10 points. Discuss strategies for winning the game.

Place Value Game

Each player will need:

- ten pieces of paper each with a digit (0, 1, 2, 3, 4, 5, 6, 7, 8, 9),
- a bag to hold the pieces of paper, and
- the following game layout for each round.

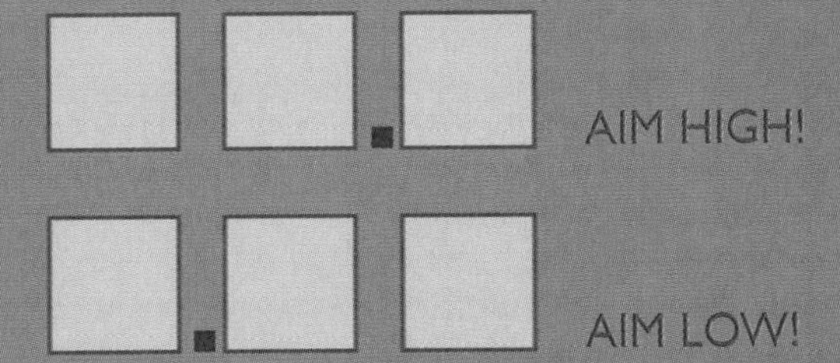

For each round of the game, players take turns drawing single digits from their own bags. After drawing a digit, the player places it in a blank square. The players take turns drawing and placing digits until they have each filled all six blanks in the layout. Once played, the digit's placement cannot be changed. Digits are not returned to the bag. The player with the greatest "Aim High" number and the player with the least "Aim Low" number each earn a point. The first player to get 10 points wins.

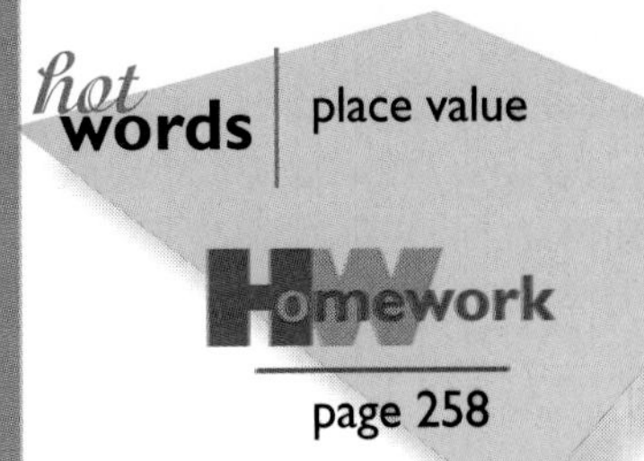

4 Get It Close Enough

ROUNDING DECIMALS

In everyday life, estimates can sometimes be more useful than exact numbers. To make good estimates, it is important to know how to round decimals. Fortunately, rounding decimals is similar to rounding whole numbers.

Round Decimals

How can you round decimals?

You can use a number line to help you round decimals.

1 Copy the number line below.

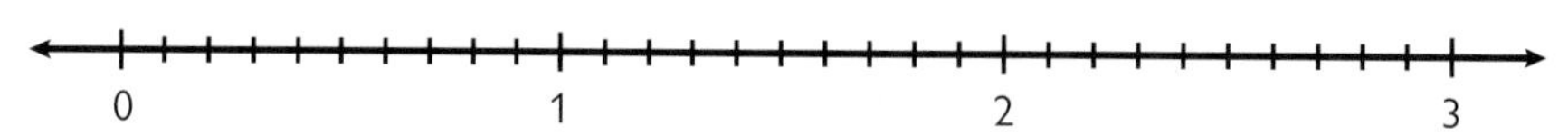

a. Graph 2.7 on the number line.

b. Is the point you graphed closer to 2 or 3?

c. Round 2.7 to the nearest whole number.

d. Use the number line to round 1.3 to the nearest whole number.

Rounding to Tenths

If the hundredths digit of the number is 4 or less, "round down" by eliminating the hundredths digit and all the digits to its right.

If the hundredths digit is 5 or more, "round up" by adding one to the tenths digit. Then, eliminate the hundredths digit and every digit to its right.

2 Round each number to the nearest tenth.

a. 9.3721 b. 6.5501 c. 19.8397

3 Round 3.1415928 to the nearest hundredth, thousandth, and ten-thousandth.

Order Fractions and Decimals

How can you order fractions and decimals?

Consider the ten numbers below.

$\frac{3}{10}$	0.75	0.125	$\frac{3}{4}$	
1.25	$\frac{75}{5}$	0.13	$3\frac{1}{2}$	$\frac{1}{100}$
0.03				

1 Write each of the ten numbers on small pieces of paper so you can move them around. Use any strategy that you find helpful to arrange these numbers in order from least to greatest. If two numbers are equal, place an equals sign between them. As you work, write notes that you can use to convince someone that your order is correct.

2 Once you are satisfied with your order, write the numbers from least to greatest. Place either $<$ or $=$ between the numbers. Make sure that your list contains all ten numbers and that each number is expressed as it was originally shown.

Write about Ordering Numbers

The numbers $\frac{19}{8}$, 2.3, and $2\frac{1}{8}$ are written in different forms. Explain how to determine which is least and which is greatest.

PHASE **TWO**

In this phase, you will be making sense of decimal operations. Deciding where the decimal point goes when you add, subtract, multiply, or divide decimals is important. Think about where you see decimals in your everyday experience. You will investigate ways to use mental math, estimations, and predictions to determine answers quickly.

Computing with Decimals

WHAT'S THE MATH?

Investigations in this section focus on:

COMPUTATION

- Understanding decimals
- Adding, subtracting, multiplying, and dividing decimals

ESTIMATION

- Using estimation to determine where to place the decimal point
- Using estimation to check whether the result makes sense

5 Place the Point

ADDING AND SUBTRACTING DECIMALS

You can use what you have learned about fractions and decimals to add decimals. One way to add decimals is to change the addends to mixed numbers and add them. Then, you can change the sum back to a decimal. In this lesson, you will use this method to develop another rule for adding decimals.

Add Decimals

How can you add decimal numbers?

Use what you know about adding mixed numbers.

1 Carl is trying to add 14.5 and 1.25, but he is not sure about the answer. He thinks his answer could be 0.270, 2.70, 27.0 or 270.

$$\begin{array}{r} 14.5 \\ +\ 1.25 \\ \hline 270 \end{array}$$

a. Use estimation to show that none of Carl's possible answers is correct.

b. Write 14.5 and 1.25 as mixed numbers. Then, add the mixed numbers.

c. Write the sum in part **b** as a decimal.

d. Explain to Carl how to add decimals.

2 Carl's sister Vonda told Carl that he should line up the decimal points. Is she correct? To find out, convert the decimals to mixed numbers, add the mixed numbers, and write the sum as a decimal. Then, use Vonda's method of lining up the decimal points and compare the sums.

a. 12.5 + 4.3 b. 22.25 + 6.4 c. 5.65 + 18.7

Why Do We Line Up the Decimal Points?

When you add decimals, you want to be sure that tenths are added to tenths, hundredths are added to hundredths, and so on.

$$\begin{array}{r} 2.35 \\ +\ 12.4 \\ \hline 14.75 \end{array}$$

	2 ones 3 tenths 5 hundredths
+	1 ten 2 ones 4 tenths
	1 ten 4 ones 7 tenths 5 hundredths

Play the "Place Value" Game

How can you create the greatest sum or difference?

Put ten pieces of paper, numbered 0, 1, 2, 3, 4, 5, 6, 7, 8, and 9, in a bag. You and your partner are going to take turns removing 6 pieces of paper, one at a time, from the bag. Each time you or your partner remove a piece of paper, write the number in one of the squares in a layout like the one below. Then, put the piece of paper back into the bag. The player with the greater sum earns a point. Play nine rounds to determine the winner. Then, answer the following questions.

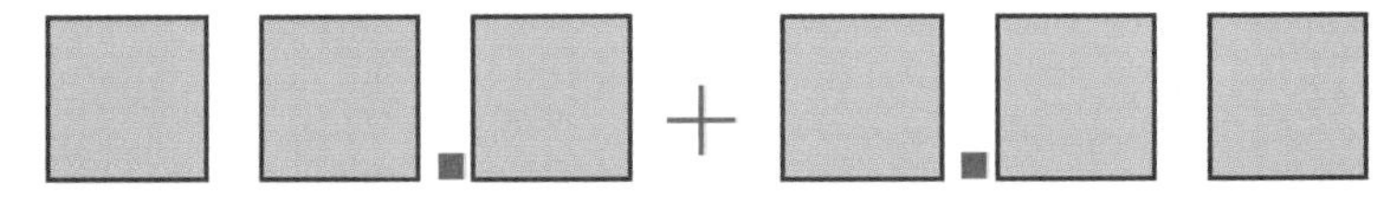

1 If the six numbers you drew were 1, 3, 4, 6, 7, and 9, how could you place the numbers to get the greatest sum?

2 How could you place the digits if you want to get the greatest difference of the two numbers?

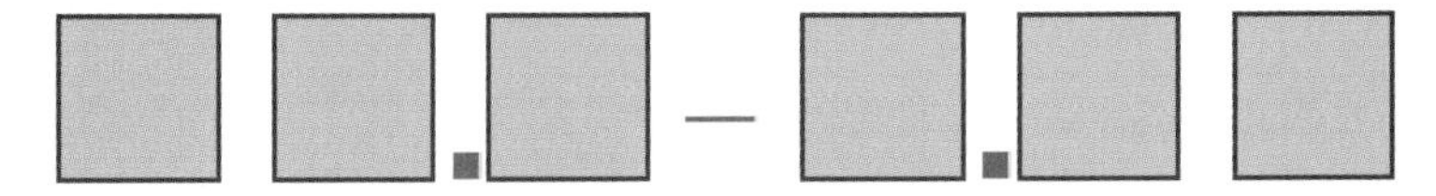

3 Explain your strategy for each "Place Value" game.

What if You Do Not Have Enough Tenths?

When trying to compute 23.25 − 8.64, you line up the decimals as you did in addition. What do you do since there are more tenths in 8.64 than in 23.25? Here is how Vonda did it. Explain what she did. Is her answer correct?

$$\begin{array}{r} \scriptstyle 1\ 12\ 12 \\ \not{2}\not{3}.\not{2}5 \\ -\ 8.64 \\ \hline 14.61 \end{array}$$

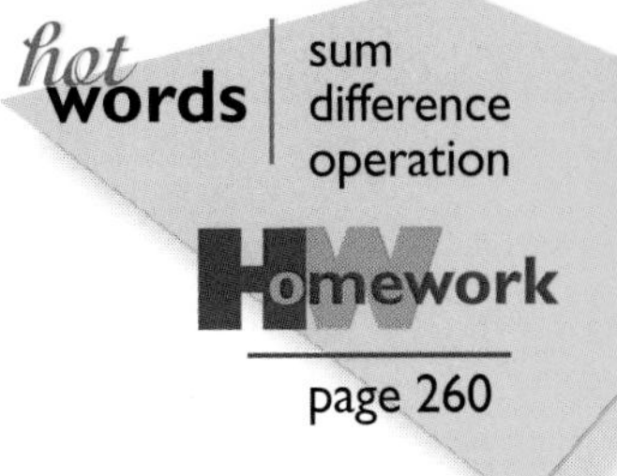

hot words
sum
difference
operation
Homework
page 260

6 More to the Point

MULTIPLYING WHOLE NUMERS BY DECIMALS

Can you multiply decimals like you multiply whole numbers? In the last lesson, you found a way to add and subtract decimals that is very similar to adding and subtracting whole numbers. In this lesson, you will learn to multiply a whole number by a decimal.

Look for a Pattern

Where do you place the decimal point?

You can use a calculator to find patterns when multiplying by special numbers.

1 Find the products for each set of problems.

a. 124.37×10
124.37×100
$124.37 \times 1{,}000$

b. 14.352×10
14.352×100
$14.352 \times 1{,}000$

c. 0.568×10
0.568×100
$0.568 \times 1{,}000$

2 Describe what happens to the decimal point when you multiply by 10, 100, and 1,000.

3 Find the products for each set of problems.

a. 532×0.1
532×0.01
532×0.001

b. $3{,}467 \times 0.1$
$3{,}467 \times 0.01$
$3{,}467 \times 0.001$

c. 72×0.1
72×0.01
72×0.001

4 Describe what happens to the decimal point when you multiply by 0.1, 0.01, and 0.001.

5 Use the patterns that you observed to find each product. Then, use a calculator to check your answers.

a. $4.83 \times 1{,}000$

b. 3.6×100

c. 477×0.01

d. 572×0.001

e. 91×0.1

f. $124.37 \times 10{,}000$

Multiply Whole Numbers by Decimals

How can you multiply a whole number by a decimal?

You can continue studying patterns to discover how to multiply by decimals.

1 Use a calculator to find the products for each set of problems.

a. 300×52
300×5.2
300×0.52
300×0.052

b. 230×25
230×2.5
230×0.25
230×0.025

c. 14×125
14×12.5
14×1.25
14×0.125

d. 18×24
18×2.4
18×0.24
18×0.024

e. 261×32
261×3.2
261×0.32
261×0.032

f. $4{,}300 \times 131$
$4{,}300 \times 13.1$
$4{,}300 \times 1.31$
$4{,}300 \times 0.131$

2 Study each set of problems. How are the products in each set similar? How are they different?

3 Predict the value of each product. Then, use a calculator to check your answers.

a. 700×2.1

b. 700×0.21

c. 14×1.5

d. 34×0.25

e. 710×0.021

f. 57×0.123

Write about Multiplication by Decimals

Reflect on the problems on this page. How did you determine where to place the decimal point after computing the whole number product? Why does this make sense?

page 261

7 Decimal Pinpoint

MULTIPLYING DECIMALS

When you multiply decimals, you can multiply whole numbers and then place the decimal in the correct place. In this lesson, you will learn how to find the product of two decimals.

Multiply Decimals

How can you multiply a decimal by a decimal?

The placement of the decimal point is important in decimal multiplication.

1 The computations for the multiplication problems are given. However, the decimal point has not been placed in the product. Use estimation to determine the location of each decimal point. Use a calculator to check your answer.

a.
$$\begin{array}{r} 25.3 \\ \times\ 1.25 \\ \hline 1265 \\ 506 \\ 253 \\ \hline 31625 \end{array}$$

b.
$$\begin{array}{r} 64.1 \\ \times\ 0.252 \\ \hline 1282 \\ 3205 \\ 1282 \\ \hline 161532 \end{array}$$

c.
$$\begin{array}{r} 1.055 \\ \times\ 4.52 \\ \hline 2110 \\ 5275 \\ 4220 \\ \hline 476860 \end{array}$$

2 How is the number of decimal places in the two factors related to the number of decimal places in the corresponding product?

3 Explain your method for placing the decimal point in the product of two decimal factors.

4 Use your method to find each product. Then, use a calculator to check your answer.

a. 12.5×2.4 **b.** 24.3×1.15 **c.** 8.9×0.003

Play another Version of the "Place Value" Game

How can you create the greatest product?

As in previous "Place Value" games, this is a game for two players. Each player should make a layout as shown below.

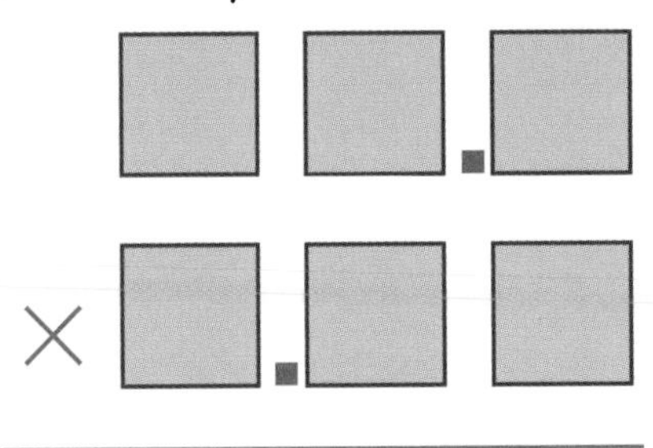

The players take turns rolling a number cube and recording the result of the roll in one of their squares. When each player has filled all of his or her squares, the players compute their products. The player with the greatest product earns a point. Play nine rounds to determine the winner. Then, answer the following questions.

1 What are the greatest and least possible products you could have while playing the game?

2 If you roll 1, 6, 5, 6, 2, and 3, what do you think is the greatest possible product? Compare your answers with those of your classmates.

Why Does It Work?

32.4		324		32.4	←		1 decimal place
× 1.23	⇒	× 123	⇒	× 1.23	←	+	2 decimal places
?????		39852		39.852	←		3 decimal places

When you compute the product of 32.4 and 1.23, you first ignore the decimal points and multiply the whole numbers 324 and 123.

When you ignore the decimal in 32.4, 324 is 10 times greater than 32.4. When you ignore the decimal in 1.23, 123 is 100 times greater than 1.23. So, the product of the whole numbers is 10 times 100, or 1,000, times the product of 32.4 and 1.23. That means that the decimal point is three places too far to the right. You must move the decimal point three places to the left to get the correct answer.

8 Patterns and Predictions

DIVIDING DECIMALS

There are many ways to think about dividing decimals. All of these methods involve things you already know how to do.

How do you determine which division problems have the same quotient?

Write Equivalent Division Problems

You can use a calculator to find patterns when dividing decimals.

1 Find the quotient for each set of problems.

a. 24 ÷ 12
2.4 ÷ 1.2
0.24 ÷ 0.12

b. 875 ÷ 125
87.5 ÷ 12.5
8.75 ÷ 1.25

c. 720 ÷ 144
72 ÷ 14.4
7.2 ÷ 1.44

2 What do you notice about the quotients for each set of division problems? How are the problems in each set similar? How they different?

3 Use a calculator to find each quotient. Then study the results.

a. 495 ÷ 33 **b.** 4,950 ÷ 330 **c.** 49,500 ÷ 3,300

d. 49.5 ÷ 3.3 **e.** 4.95 ÷ 0.33 **f.** 0.495 ÷ 0.033

4 Which of the following quotients do you think will be equal to 408 ÷ 2.4? Explain your reasoning. Then, use a calculator to check your predictions.

a. 4,080 ÷ 24 **b.** 40,800 ÷ 240 **c.** 408 ÷ 24

d. 40.8 ÷ 0.24 **e.** 40.8 ÷ 2.4 **f.** 4.08 ÷ 0.024

5 Rewrite each division problem as an equivalent problem involving whole numbers. Then, find the quotient.

a. $1.2\overline{)36}$ **b.** $1.25\overline{)22.5}$ **c.** $1.45\overline{)261}$

Divide Decimals

What happens if the quotient is a decimal?

If you compute $12 \div 5$, you would get $\frac{12}{5}$ or $2\frac{2}{5}$. Notice that $2\frac{2}{5}$ is the same as $2\frac{4}{10}$ or 2.4. You can get this same result as follows.

```
   2.4
5)12.0
  10
   2.0
   2.0
     0
```

Notice that 2.0 is equivalent to 20 tenths. Since $20 \div 5 = 4$, you know that 20 tenths divided by 5 is 4 tenths.

1 Work with a classmate to help each other understand that $19 \div 4 = 4.75$. Two ways of displaying the calculation are shown below.

```
   4.75          4.75
4)19.00       4)19.00
  16            16
   3.0           30
   2.8           28
   0.20           20
   0.20           20
   0.00            0
```

2 Copy and complete the division problem below. Then, check your answer by multiplying and by using a calculator.

```
   32.1
8)257.000
  24
   17
   16
    10
     8
     2
```

3 Find each quotient. Then, use a calculator to check your answer.

a. $326.7 \div 13.5$ b. $323.15 \div 12.5$ c. $428.571 \div 23.4$

Write about Dividing Decimals

Write an explanation of how to divide decimals.

It Keeps Going and Going

WRITING
REPEATING AND
TERMINATING
DECIMALS

You have already learned to use division to change fractions to decimals. In previous lessons, the decimals could be represented using tenths, hundredths, or thousandths. In this lesson, you will discover decimals that never end.

Write Repeating Decimals

How far can decimals go?

When you use a calculator to find $4 \div 9$, you may get 0.4444444 or 0.44444444 or 0.44444444444. The answer will depend on the number of digits your calculator shows. What is the *real* answer?

1. Find the first five decimal places in the decimal representation of $\frac{4}{9}$ by dividing 4.00000 by 9. Show your work.
2. When Vonda did step **1**, she said, "The 4s will go on forever!" Is she correct? Examine your division and explain.
3. Another repeating decimal is the decimal representation of $\frac{5}{27}$. What part is repeating? How do you know it repeats forever?
4. Every fraction has a decimal representation that either repeats or terminates. Find a decimal representation for each fraction by dividing. Tell if the decimal is *repeating* or *terminating*.

   a. $\frac{5}{16}$ **b.** $\frac{4}{27}$ **c.** $\frac{1}{6}$ **d.** $\frac{4}{13}$ **e.** $\frac{13}{25}$

Repeating Decimals

The decimal equivalent for $\frac{2}{9}$ is 0.2222. . . .This is called a **repeating decimal** because the 2 repeats forever. For 0.3121212 . . . , the 12 repeats. For 0.602360236023. . . , the 6023 repeats. Sometimes a repeating decimal is written with a bar over the part that repeats.

$0.\overline{2}$ $0.3\overline{12}$ $0.\overline{6023}$

A decimal that has a finite number of digits is called a **terminating decimal.**

Use Decimals

What is Celsius?

The United States uses the Fahrenheit scale to measure temperature. Most other parts of the world use the Celsius scale. To change degrees Celsius to degrees Fahrenheit, you can use the following two steps.

- Multiply the Celsius temperature by 1.8.
- Add 32 to the product.

1 Copy the table below. Complete the table by using the two steps.

°C	0	5	7	9	12	15	20	30	40
°F	32								

2 The two steps give an exact conversion. To *approximate* the Fahrenheit temperature given a Celsius temperature, you can double the Celsius temperature and then add 30.

Use this method to approximate each Fahrenheit temperature in the table. Compare the results with the actual temperatures. Which of these temperatures is within 2 degrees of the actual temperature?

3 Work with one or more of your classmates to devise a way to change a Fahrenheit temperature to a Celsius temperature. Use the table in step **1** to check your method. Then, use your method to change each Fahrenheit temperature to Celsius.

a. 50°F **b.** 95°F **c.** 64.4°F

Write about Temperature Conversion

Teresa and Donte's math assignment was to keep track of the daily high temperatures for three days and then find the average high temperature in Celsius. Teresa and Donte only have a thermometer that measures degrees Fahrenheit. They recorded the following temperatures.

Friday: 68°F Saturday: 56°F Sunday: 74°F

Teresa said that they need to average the three temperatures first and then convert to Celsius. Donte said that they could convert all three temperatures to Celsius and then find the average.

- Will both methods work?
- Will one be more accurate than the other?
- Find the average temperature using both methods. Explain which way you think is more efficient and why.

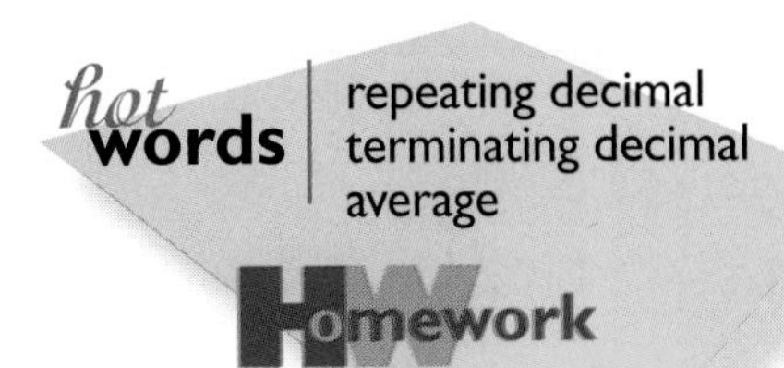

page 264

PHASE THREE

Percents are used everyday. In this phase, you will use mental math and estimation to calculate percents of a number. You will explore the relationship between fractions, decimals, and percents and study how percents are used and misused in the world around us.

Percents

WHAT'S THE MATH?

Investigations in this section focus on:

COMPUTATION

- Understanding the relationships among fractions, decimals, and percents
- Using mental math to get exact answers with percents
- Calculating percent of a number

ESTIMATION

- Using estimation to check whether results are reasonable

10 Moving to Percents

WRITING EQUIVALENT FRACTIONS, DECIMALS, AND PERCENTS

When you express a number using hundredths, you can also use percents. The word *percent* means hundredths. The symbol % is used to show the number is a percent.

How can you determine a percent of a whole?

Interpret Percents

Twenty-three hundredths of a whole is twenty-three percent of a whole.

$$\frac{23}{100} = 23\%$$

For each drawing, determine what percent of the whole is shaded.

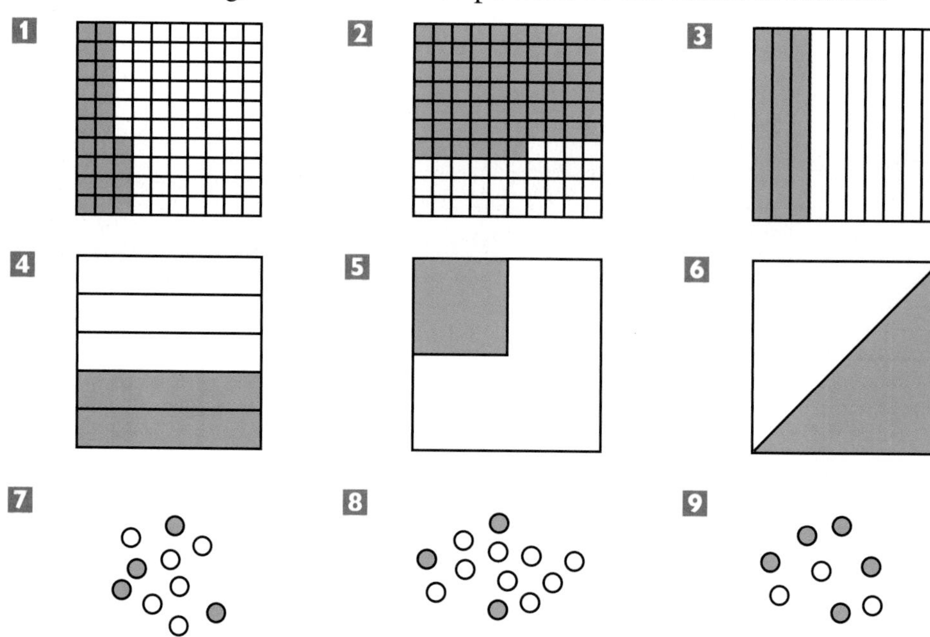

10 Today, the United States has 50 states. Only 13 of the states took part in signing the Declaration of Independence in 1776. What percent of today's states signed the Declaration of Independence?

Write Different Forms for Parts of a Whole

How are fractions, decimals, and percents related?

Here are some examples showing how you can write a percent as a fraction or decimal, or write a fraction or decimal as a percent.

41% can be written as $\frac{41}{100}$ and as 0.41.

0.09 can be written as $\frac{9}{100}$ and as 9%.

$\frac{4}{5}$ can be written as $\frac{8}{10}$, as 0.8, and as 80%.

1 Each of the following numbers can be written as a fraction, as a decimal, and as a percent. Write each number in two other ways.

a. $\frac{3}{10}$ **b.** 80% **c.** 0.75

d. 55% **e.** 0.72 **f.** $\frac{7}{25}$

2 Stephanie collects decks of playing cards. She has 200 decks, and 35 decks are from foreign countries.

a. Use a grid of 100 squares to represent her collection. How many decks does each square represent?

b. Shade the appropriate number of squares to represent the 35 decks from foreign countries.

c. Write a fraction, a decimal, and a percent that represent the part of her collection that are from foreign countries.

3 Write each number in two other ways.

a. 0.755 **b.** 5.9% **c.** $\frac{3}{8}$

d. $\frac{5}{16}$ **e.** 0.003 **f.** 17.4%

Summarize the Conversions

Write instructions someone else can follow to perform each conversion.

- Change a percent to an equivalent decimal.
- Change a percent to an equivalent fraction.
- Change a fraction to an equivalent percent.
- Change a decimal to an equivalent percent.

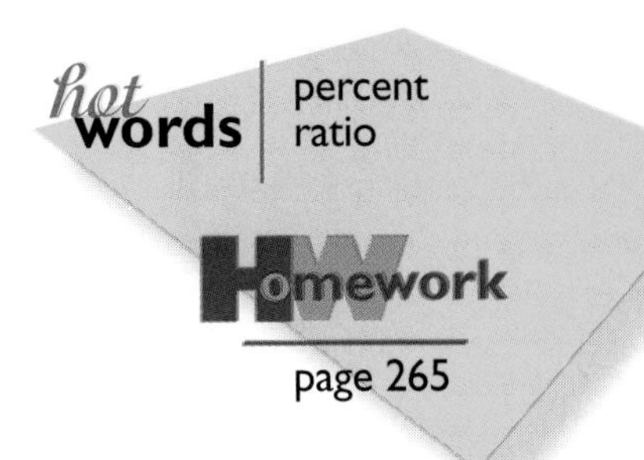

11 Working with Common Percents

REPRESENTING COMMON PERCENTS

Amaze your friends! Impress your parents! With knowledge of just a few percents, you can determine reasonable answers.

How do common fractions and decimals relate to percents?

Model Important Percents

You often see fractions such as $\frac{1}{2}$, $\frac{1}{3}$, $\frac{1}{4}$, and $\frac{3}{4}$. You probably have a good sense of what these numbers mean.

1 Consider the fractions $\frac{1}{2}, \frac{1}{3}, \frac{2}{3}, \frac{1}{4}, \frac{3}{4}, \frac{1}{5}, \frac{2}{5}, \frac{3}{5}, \frac{4}{5}$.

a. Represent each fraction on a 10×10 grid or on a number line. Label each grid or point with the fraction it represents.

b. Label each grid or point with the equivalent decimal and equivalent percent.

c. Represent 0 and 1 with grids or on the number line.

2 For each percent, make a new grid or add a new point to the number line. Label each new grid or point with the percent, fraction, and decimal it represents.

a. 10% b. 30% c. 70% d. 90%

3 Which fraction from step **1** is closest to each percent?

a. 15% b. 85% c. 39% d. 73%

Estimating with Percents

79% is between $\frac{3}{4}$ or 75% and $\frac{4}{5}$ or 80%. It is closer to $\frac{4}{5}$ than $\frac{3}{4}$.

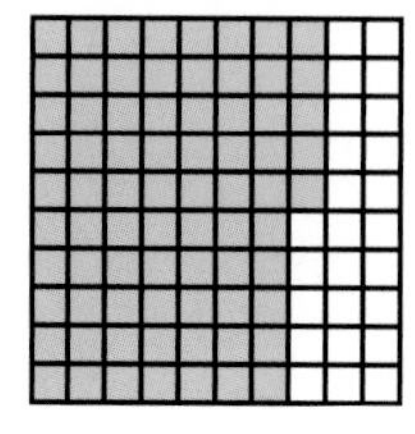

$\frac{3}{4}$ or 75%

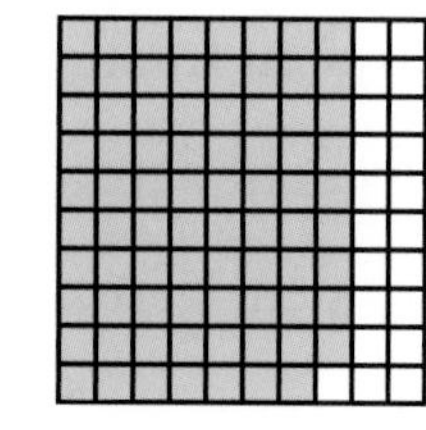

79%

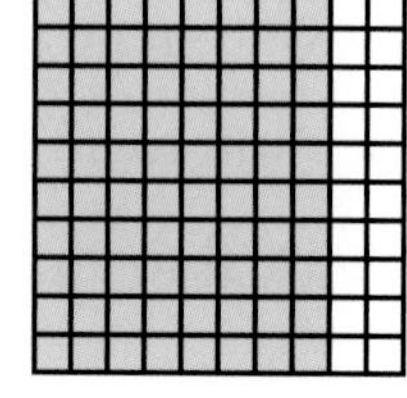

$\frac{4}{5}$ or 80%

Find Important Percents

Three important percent values are 50%, 10%, and 1%. Fortunately, these are percents that you can calculate quickly.

1 Use the numbers in List A that you compiled in class.

a. Choose a number. On your own, find 50%, 10%, and 1% of the number. Use any strategy that makes sense to you. Record each answer and explain how you found it.

b. On your own, find 50%, 10%, and 1% of another number. See if you can use strategies that are different than those used to find percents of the first number. Look for patterns in your answers.

c. Share your answers and strategies with a partner. Discuss any patterns you see.

d. With your partner, find 50%, 10%, and 1% of each number not already used by you or your partner. Remember to record your answers and strategies.

2 With your partner, choose at least three numbers from List B. Find 50%, 10%, and 1% of each number.

How can you use mental math to find 50%, 10%, and 1% of a number?

Write about Percents

Write a short paragraph explaining how you can quickly find 50%, 10%, and 1% of any number.

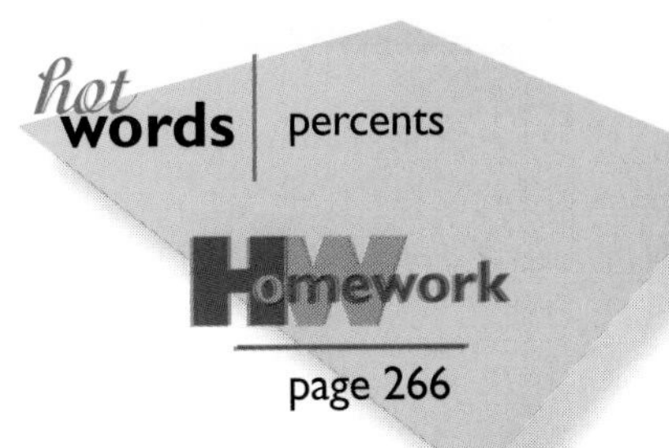

12 Percent Power

CALCULATING PERCENTS

How can you find the percent of a number? In Lesson 11, you found 50%, 10%, and 1% of a number. In this lesson, you will find any percent of a number.

How can you estimate and find the actual percent of a number?

Find a Percent of a Number

In this investigation, you will use fractions to estimate a percent of a number. Suppose you want to estimate 58% of 40.

58% is a little less than 60% or $\frac{3}{5}$.

$\frac{3}{5}$ of 40 is 24.

So, 58% of 40 is a little less than 24.

1 For each problem, find a fraction that you think is close to the percent. Use the fraction to estimate the quantity. Record your answers in the "Estimate" column of the handout Finding Any Percent.

a. 63% of 70

b. 34% of 93

c. 12% of 530

d. 77% of 1,084

2 Complete the first four rows of the handout. Check the exact value by comparing it to the estimate you found in step **1**.

3 Choose six numbers and six percents from the following list. Write the numbers in the proper columns of the handout. Use at least one three-digit number and at least one four-digit number. Then, complete the table.

Number				Percent Wanted			
25	40	36	90	15%	18%	90%	3%
390	784	135	124	38%	23%	44%	83%
9,300	4,220	3,052	4,040	9%	51%	73%	62%

Use One Hundred to Find Percents

Liz and her twin brother Nate like trading cards. Some are sports cards (baseball, football, and auto racing), some are comic book cards, and some are game cards. Together, they have 1,000 cards. Liz has separated the cards into 10 sets of 100, so that each set has the same number of each type of card.

As you answer each question, think about how you determined your answer.

1 Each set has 10 auto racing cards. What percent of a set is auto racing cards? What percent of the 1,000 cards is auto racing cards?

2 For each set, 23% are comic book cards. How many comic book cards are in each set? How many comic book cards do Liz and Nate have in all?

3 Of the 1,000 cards, 340 are baseball cards. What percent of the 1,000 cards is baseball cards? What percent of each set is baseball cards?

4 There are 8 football cards in each set of 100, so 8% of each set are football cards. Nate combined five sets into a group of 500, so he could have half of all the cards.

a. How can you find the number of football cards in each set of 500 cards? in all 1,000 cards?

b. How can you find the percent of football cards in each set of 500 cards? in all 1,000 cards?

5 Liz still has sets of 100. In each set, there are 25 game cards.

a. How many sets would she need to combine to have exactly 100 game cards?

b. Of the 25 game cards in each set, 12 cards are one type of game. How many would there be in the combined 100 game cards? What percent of the 100 game cards represents this game?

How can you calculate percents using sets of 100?

Write about Using Proportions

When two sets have the same percent or fraction of one item, then the sets are *proportional.* For example, the fraction of baseball cards in each small set is the same as the fraction of baseball cards in the big set. Write a paragraph explaining how to use the idea of proportions to find the percent of a number.

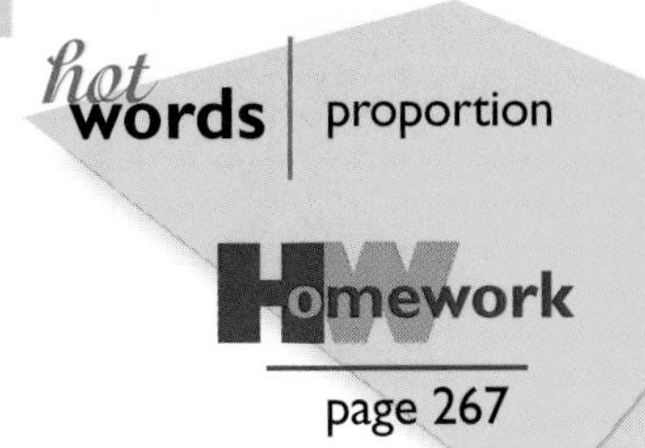

13 Less Common Percents

UNDERSTANDING PERCENTS LESS THAN 1 AND GREATER THAN 100

Sometimes percents can represent a very small part of a whole or a part greater than a whole. In this lesson, you will study these types of percents.

Understand Percents Less than One

What does a percent less than one mean?

Some percents are less than 1%. To understand the meaning of these types of percents, answer the following questions.

1 Carlos won a gift certificate for $200 at an electronics store. He used his gift certificate to purchase the six items listed below. What percent of the gift certificate was used for each item?

a. a game system: $100
b. a game cartridge: $30
c. another game cartridge: $20
d. a DVD movie: $24
e. a movie soundtrack: $12
f. a rock CD: $13

2 Add the prices in step **1**.

a. How much of the gift certificate remained unspent? What percent of the $200 was left unspent?

b. If all $200 had been spent, what percent of the gift certificate was spent?

3 Write an equivalent decimal and an equivalent fraction for each percent.

a. 0.2% **b.** 0.7% **c.** $0.\overline{3}\%$ **d.** $\frac{1}{10}\%$ **e.** $\frac{2}{5}\%$

4 Which of the following numbers represents a percent less than one? Try to answer without converting the fraction or decimal to a percent. Be prepared to discuss your answers.

a. 0.1 **b.** 0.004 **c.** 0.013 **d.** 0.0082
e. $\frac{1}{200}$ **f.** $\frac{3}{200}$ **g.** $\frac{1}{34}$ **h.** $\frac{10}{1,000}$

Understand Percents Greater than One Hundred

What does a percent greater than 100 mean?

Nichelle is organizing her family's collection of photographs. She has several photo albums that are the same size.

1 Nichelle filled $\frac{3}{4}$ of one photo album with the first box of photographs. The following model represents $\frac{3}{4}$ of the album.

What percent of the album was used?

2 The next box had the same number of photographs in it. The space required for both boxes is twice as much as the first box.

a. Draw a model similar to the one in step **1** to represent how much of the photo albums are filled.

b. What fraction of one album does your drawing represent?

c. What percent of one album does your drawing represent? How did you determine your answer?

3 Each album held 60 photographs.

a. Find the number of photographs in the first box.

b. Find the number of photographs in the second box.

c. When Nichelle was finished, she had filled three albums and $\frac{1}{5}$ of another album. What percent of one album did she fill? How many photographs does Nichelle have?

Make Models

Make a model to represent $\frac{1}{2}$%. Then, make a model to represent 250%. Explain why each model represents the appropriate percent.

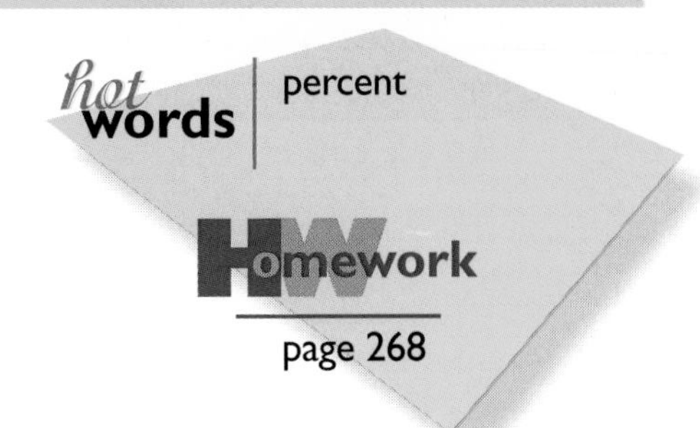

14 Give It to Me Straight

UNDERSTANDING USES AND MISUSES OF PERCENTS

Sometimes percents are misused in everyday life. Misleading percents may seem correct at first. However, if you use what you know about percents, you will quickly discover which statements are wrong.

Examine Uses of Percents

Which statements use percents correctly?

Study the statements involving percents. For each statement, decide if the percent is used correctly.

- If a statement has been used correctly, describe what the statement means. In some cases, you may want to make up some numbers to give an example of what it means.
- If a statement has not been used correctly, explain why it is incorrect.

1. Jewel said, "I have finished about 50% of my homework."
2. Gracie's Department Store is offering 15% off the regular price on all men's shoes.
3. Of the people who are older than 60 years old and who are living alone, 34% are women and only 15% are men.
4. Central High School enrollment is 105% of last year's enrollment.
5. Ice cream gets its texture by adding air. A few ice cream makers double the volume by having 100% air.
6. One brand of snacks has 13% of the $900 million annual sales for similar snacks.
7. Teenagers represent 130% of the attendance at an amusement park.
8. The rainfall in a particular area is 120% less than it was 25 years ago.

Navigate the Maze

For this maze, you want to create a path of marked squares that lead from the lower left corner to the upper right corner using the following rules.

- The lower left square is marked first.
- You can mark an unmarked square to the left or right or above or below the last square marked.
- You can only mark the square whose value is closest to the value in the last square marked.

In the example below, the last square marked was 45%. There are three possibilities for the next square, $\frac{4}{10}$, $\frac{4}{5}$, or $\frac{1}{4}$. Since $\frac{4}{10}$ is closest to 45%, that square must be marked next.

17%	$\frac{4}{5}$	$0.\overline{3}$
$\frac{4}{10}$	45% (marked)	$\frac{1}{4}$
0.2 (marked)	0.3 (marked)	$\frac{3}{4}$

Try the maze on the handout.

Can you find the correct path through the maze?

Interpret the Data

Trent and Marika surveyed 150 students in their school. The results are shown below.

Percent of Students Who Own Various Types of Pets

Dogs: $33.\overline{3}$%   Cats: 30%   Birds: 6%   Other: 2%   None: 44%

Total Number of Pets Owned by the Students

Dogs: 86   Cats: 95   Birds: 14   Other: 5

1 How many students own each type of pet? How many students own no pets?

2 What percent of all the pets are dogs? cats? birds? other pets?

3 Marika decided they had made a mistake. When she added the number of students from part **1**, she said they had too many students. Why do you think she said that? Is she correct?

percent

page 269

PHASE FOUR

Negative numbers are used when the temperature is really cold, when someone is losing a game, and when something is missing. All of the numbers you have already used have negative counterparts. If you can handle the whole numbers 0, 1, 2, 3, 4, . . . , you can learn the strategies to handle negative numbers -1, -2, -3, -4, . . . and all integers.

The Integers

WHAT'S THE MATH?

Investigations in this section focus on:

NUMBER and COMPUTATION

- Using negative integers to indicate values
- Placing negative integers on the number line
- Adding and subtracting integers
- Writing equivalent addition and subtraction problems

ALGEBRA FUNCTIONS

- Recognizing and writing equivalent expressions

15 The Other End of the Number Line

UNDERSTANDING NEGATIVE NUMBERS

What numbers are less than zero? When the meteorologist says the temperature is 5° below zero, he or she is using a number less than zero.

Play "How Cold Is It?"

How can you order numbers less than zero?

"How Cold Is It?" is a game for two people.

1 Use the following rules to play "How Cold Is It?"

- One person thinks of a cold temperature between 0°F and −50°F and writes it down.
- The other person tries to guess the temperature. After each guess, the first person will say if the guess is warmer or colder than his or her temperature.
- When a person guesses the temperature, record the number of guesses needed. Then, the players switch rolls.
- After four rounds, the player who guessed a temperature using the fewest guesses is the winner.

60
50
40
30
20
10
0
-10
-20
-30
-40
-50

2 Use the record low temperatures below to determine which city had the coldest record low temperature. Determine which city had the warmest record low temperature. List the cities in order from coldest to warmest record low temperature.

Record Low Temperatures

City	Temperature	City	Temperature
Nome, Alaska	−54°F	Missoula, Montana	−33°F
Phoenix, Arizona	17°F	El Paso, Texas	−8°F
Apalachicola, Florida	9°F	Burlington, Vermont	−30°F
St. Cloud, Minnesota	−43°F	Elkins, West Virginia	−24°F

Play "Who's Got My Number?"

Play "Who's Got My Number?" with the whole class or with a small group.

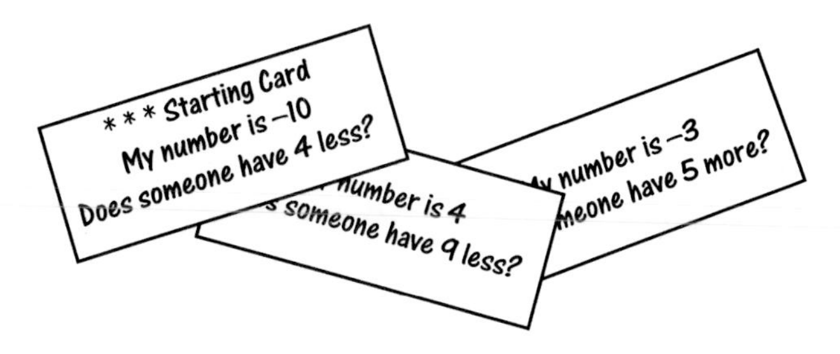

* * * Starting Card
My number is −10
Does someone have 4 less?
number is 4
someone have 9 less?
number is −3
meone have 5 more?

How can you show numbers less than zero on a number line?

Who's Got My Number?

To play the game, your teacher will give your group a set of cards and a number line recording sheet.

- Shuffle and deal out all of the cards.
- The person with the starred starting card begins by reading the card.
- Locate the number on the number line. Players should use the number line and the clue to determine the next number.
- The person who has that number on one of his or her cards reads it and play continues.

Make Your Own Set of Clues

Write a set of your own clues for the game you just played.

- Include clues so that the numbers go as low as -15 and as high as 10.
- Write at least six clues in your set.
- Place a star on the starting card.
- Draw a number line and show the moves to make sure your clues are correct.

16 Moving on the Number Line

USING NUMBER LINES TO ADD INTEGERS

If the morning temperature is −5°F and the temperature rises 20°F by afternoon, what is the afternoon temperature? To answer this question, you must add integers. In this lesson, you will use number lines to add integers.

How can you use number lines to add integers?

Add Integers on a Number Line

Adding a positive number corresponds to moving to the right.

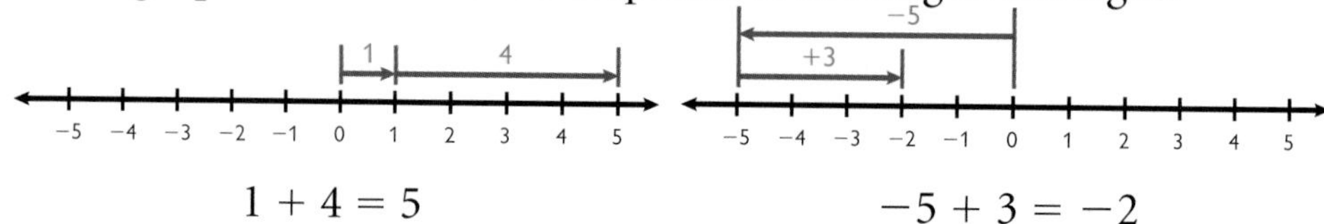

$1 + 4 = 5$ $\quad -5 + 3 = -2$

Adding a negative number corresponds to moving to the left.

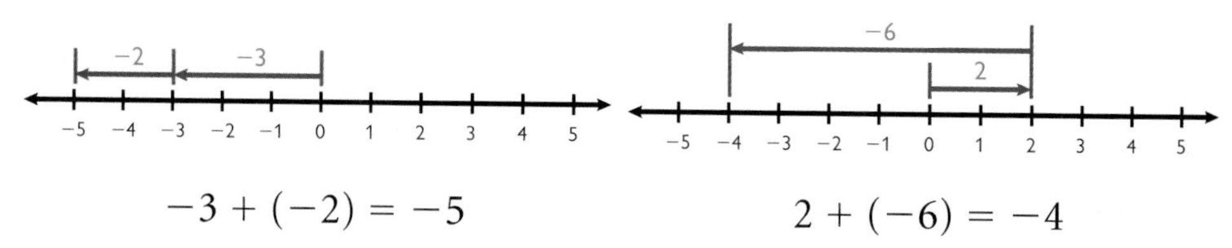

$-3 + (-2) = -5$ $\quad 2 + (-6) = -4$

Play "Number Line Showdown" with a partner.

Number Line Showdown

- Remove the face cards from a regular deck of cards. Shuffle and deal out the cards facedown in front of the players. The black cards will represent positive numbers, and the red cards will represent negative numbers.
- Player 1 turns over the top card from his or her pile. He or she draws an arrow on a number line to represent moving from 0 to the number.
- Player 2 turns over his or her top card. He or she draws an arrow on the number line to show moving from Player 1's number to complete the addition.
- If the sum is positive, Player 1 gets the cards played. If the sum is negative, Player 2 gets the cards played. If the sum is zero, no one gets the cards.
- The players take turns being Player 1 and Player 2.
- The player with the most cards after all the cards have been played wins.

Play Variations of "Number Line Showdown"

Now, you will play a more challenging variation of "Number Line Showdown". Decide which variation of the game you and your partner would like to play.

How can you add more than two integers?

Variations of Number Line Showdown

Variation 1

Player 1 turns over his or her top two cards, uses these numbers to draw an addition problem on the number line, and finds the sum. Then, Player 2 turns over two cards and does the same. The player with the greater sum gets the cards played.

Variation 2

Player 1 turns over his or her top four cards. He or she shows the addition of the four numbers on the number line. Then, Player 2 turns over four cards and does the same. The player with the greater sum gets the cards played.

Write about Addition on a Number Line

You are going to begin to create a handbook that explains how to add and subtract positive and negative numbers. Your handbook should be written for someone who does not know anything about adding and subtracting these numbers.

Today, you will write Part I. This should include a short description of adding positive and negative numbers using a number line.

- Make up three examples of addition problems. Include a positive number plus a negative number, a negative number plus a positive number, and a negative number plus a negative number.
- For each problem, show how to use the number line to solve the problem. Write an addition equation to go with each drawing.
- Include any hints or advice of your own.

page 271

17 Taking the Challenge

USING CUBES TO ADD INTEGERS

Cubes can also be used to model addition of integers. In this lesson, you will use several cubes of one color to represent positive integers and several cubes of another color to represent negative integers.

What other model can be used to represent addition of integers?

Develop another Model for Addition

To play "Color Challenge", one partner will need 10 pink cubes to represent positive numbers, and the other partner will need 10 green cubes to represent negative numbers. You will also need a spinner to share. If you and your partner are using cubes of other colors, change the names of the colors on the spinner.

Play several rounds of "Color Challenge".

Color Challenge

- Take turns spinning the spinner. After each spin, place the appropriate cubes as stated on the spinner in the center of the desk or table.
- As you play, work with your partner to develop a method for keeping track of who has more cubes in the center. How many more cubes does the player have?
- The first player to have at least 4 more cubes in the center than the other player wins.

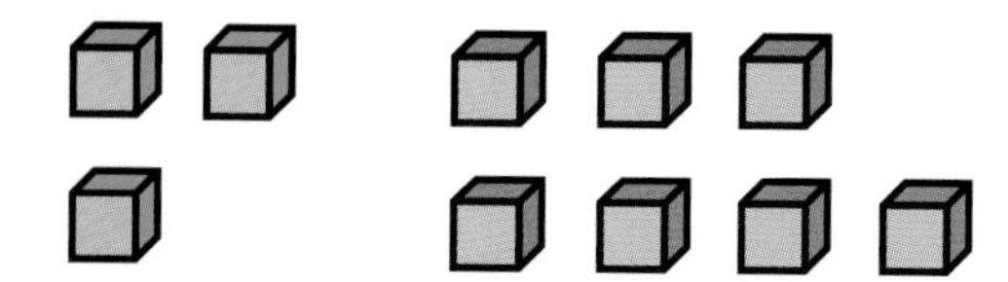

There are 4 more pink cubes than green cubes, so pink has won.

Write Addition Equations

How can you use cubes to represent addition of numbers?

Play "Addition Bingo". As in "Color Challenge", one player will need 10 cubes of one color, and the other player will need 10 cubes of another color. You will also need a new spinner and a game board.

Addition Bingo

- The first player spins the spinner twice, places the appropriate cubes on the desk, and writes the addition problem represented by the cubes.
- The first player finds the sum of the cubes by removing the same number of each color so that only one color remains. The remaining cube(s) represent the sum.
- After finding the sum, the first player crosses off a matching addition problem on his or her game board. If there is no matching addition problem or if it has already been crossed off, the player loses his or her turn.
- The players continue to take turns spinning, writing addition problems, and crossing off squares.
- The first player to have crossed off four squares in a row vertically, horizontally, or diagonally wins.

Write about Addition Using Cubes

Write Part II of your handbook about addition and subtraction of positive and negative numbers. Describe adding positive and negative numbers using cubes.

- Make up three examples of addition problems. Include a positive number plus a negative number with a positive sum, a positive number plus a negative number with a negative sum, and a negative number plus a negative number.
- For each problem, show how to use cubes to solve the problem. Write an addition equation to go with each drawing.
- Include any hints or advice of your own.

hot words | sum
zero pair

page 272

18 The Meaning of the Sign

USING NUMBER LINES TO SUBTRACT INTEGERS

Mathematical symbols, like words, can sometimes mean more than one thing. One of those symbols is the – sign. Sometimes it means to subtract and sometimes it shows that the number is negative.

How can you use number lines to subtract integers?

Subtract Integers on a Number Line

Subtracting a positive number corresponds to moving to the left on the number line.

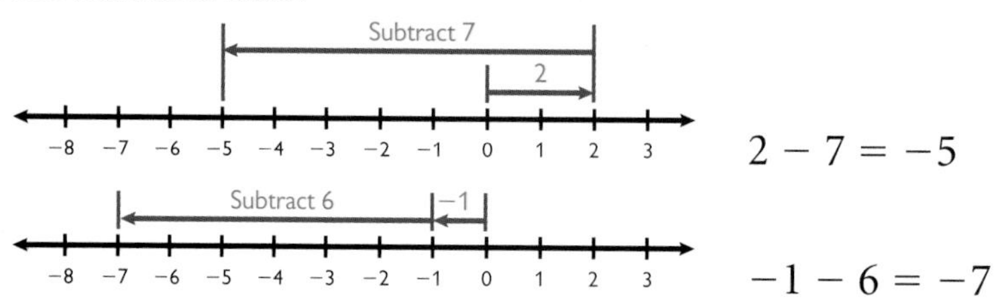

$2 - 7 = -5$

$-1 - 6 = -7$

Your teacher will give you a set of four number cubes and a recording sheet.

1. With a partner, roll the number cubes. Write the numbers in the squares on the number sheet.
2. Each student should use the four numbers and addition and subtraction signs to write an expression. Try to arrange the numbers and signs so that the result is the least possible number. Use at least one addition sign and at least one subtraction sign.
3. On a number line, use arrows to represent the entire expression in step **2**. Write the equation.
4. Write a second expression so that the result is the greatest possible number.
5. On a number line, use arrows to represent the entire expression in step **4**. Write the equation.
6. Compare your answers with your partner's answers. Who got the least answer? Who got the greatest answer?

Solve Subtraction Puzzles

How can you use what you know about subtraction to solve puzzles?

To solve the puzzles on this page, start with the number in the left column and subtract the number in the top row.

1 The example in the puzzle below shows $0 - 4 = -4$. Copy and complete the puzzle.

Second Number (top row); **First Number** (left column)

−	3	6	4	8
2			↓	
−1			↓	
0	→	→	−4	
4				

2 Copy and complete the following two puzzles.

−	2	4		
5				
−2			−3	
		−1		−5
			−7	

−	2	9		
4			−2	
−9				
			−1	
		−3		−5

Write about Subtraction on a Number Line

Write Part III of your handbook about addition and subtraction of positive and negative numbers. Describe subtraction using number lines.

- Make up three examples of subtraction problems. Include a positive number minus a positive number with a positive answer, a positive number minus a positive number with a negative answer, and a negative number minus a positive number.
- For each problem, show how to use the number line to solve the problem. Write a subtraction equation to go with each drawing.
- Include any hints or advice of your own.

19 The Cube Model

USING CUBES TO SUBTRACT INTEGERS

When you use cubes to model subtraction, sometimes you will not have enough cubes to take away. To solve these problems, you can add zero pairs. A zero pair is one positive cube and one negative cube.

How can you use cubes to subtract integers?

Subtract Integers Using Cubes

When using cubes to model integers, remember that pink cubes represent positive numbers and green cubes represent negative cubes. You may use other colors, but you must first decide which color represents positive numbers and which represents negative numbers.

1 Use cubes to find each difference.

a. $2 - 7$ **b.** $-1 - 5$ **c.** $-2 - 2$

d. $-6 - 3$ **e.** $-3 - 7$ **f.** $4 - 9$

g. $-5 - 1$ **h.** $-4 - 4$ **i.** $-3 - 5$

2 Study the subtraction problems in step **1**. Use what you have learned about subtraction to find each difference.

a. $-\frac{2}{3} - \frac{1}{3}$ **b.** $\frac{3}{4} - \frac{6}{4}$ **c.** $-4.5 - 1.4$

d. $5.2 - 10.1$ **e.** $\frac{1}{2} - \frac{7}{8}$ **f.** $-\frac{4}{5} - \frac{19}{20}$

Subtraction with Cubes

To find $-2 - 1$, use the following steps.

- Use two green cubes to represent -2.

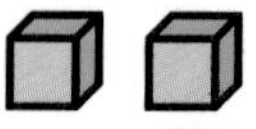

- You need to take away one pink cube, but you do not have any pink cubes.

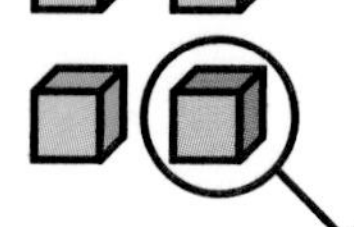

- Add one zero pair, so that you will have one pink cube to take away.
- Take away one pink cube.
- There are three green cubes left. The answer is -3.

Find the Mistakes

Can you find mistakes in subtraction?

Ms. Parabola collected her students' work and found a number of errors. Each of the following solutions has something wrong with it. Determine why it is incorrect. Write an explanation of what the student did wrong. Then, solve it correctly.

1 I solved the problem $-3 - 3$ by setting out 3 negative cubes. Then, I took away 3 negative cubes. I got 0.

2 I solved the problem $-2 - 4$ by adding two zero pairs to 2 negative cubes. I took away 4 negative cubes. I got +2.

3 I solved the problem $3 - 5$ by adding 5 positive cubes to the 3 positive cubes. Then, I took away 5 positive cubes. I got the answer +3, but that cannot be right!

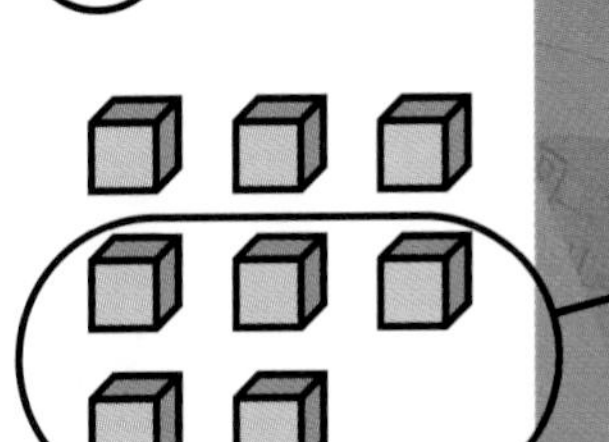

4 I solved the problem $3 - 2$ by adding 2 zero pairs to the 3 positive cubes. Then, I took away 2 negative cubes. I got +5.

Write about Subtraction Using Cubes

Write Part IV of your handbook about addition and subtraction of positive and negative numbers. Describe subtraction using cubes.

- For each of the following problems, show how to use cubes to solve the problem. Write a subtraction equation to go with each drawing.

  $5 - 8$ $\quad -4 - 2$ $\quad -1 - 3$

- Include any hints or advice of your own.

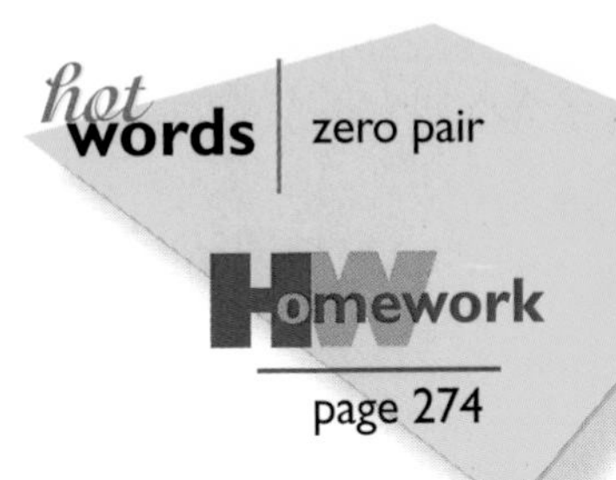

20 Write It another Way

EXPLORING EQUIVALENT STATEMENTS

You already know that addition and subtraction are related. In this lesson, you will learn how to write an addition problem as a subtraction problem and a subtraction problem as an addition problem.

Look for Patterns in Addition and Subtraction

How are addition and subtraction related?

Chris made the following list of addition equations. The first number remains the same. The second number is always one less than the number before it.

$4 + 3 = 7$
$4 + 2 = 6$
$4 + 1 = 5$
$4 + 0 = 4$
$4 + (^-1) = 3$
$4 + (^-2) = 2$
$4 + (^-3) = 1$
$4 + (^-4) = 0$
$4 + (^-5) = {}^-1$
$4 + (^-6) = {}^-2$

1. Make a similar list of subtraction equations. Start with $4 - 10$. Make the second number one less than the number before it. Stop when the second number is zero.

2. Compare the two lists. Which problems have the same answers? Describe any patterns you notice.

3. Use what you notice about the patterns to write $3 + (-7)$ as a subtraction problem. Use number cubes to show how you would find the answer for each problem.

4. Write $5 - 8$ as an addition problem. Use a number line to show how you would find the answer for each problem.

Write a Test

Here is your chance! You are going to write a test. Your test should not be so easy that everyone will get all the answers right. It should not be so hard that no one can answer the questions. Be sure to include at least one of each of the following types of problems.

- an integer addition problem with a negative answer
- an integer addition problem with a positive answer
- an integer subtraction problem with a negative answer
- an integer subtraction problem with a positive answer
- a true/false problem about estimating an answer to either an addition or subtraction problem (You may use whole numbers, fractions, or decimals. Make sure the problem shows you whether the test taker understands the order of positive and negative numbers on the number line.)
- a multiple-choice problem that uses the order of operation rules (You must include positive and negative numbers and four answer choices.)
- a writing problem that asks the test taker to tell about a new thing he or she learned about adding and subtracting positive and negative numbers (You might ask the test taker to answer a question, to write an explanation for someone, or to respond to a Dr. Math letter that you create.)

Prepare an answer key for your test. Preparing the answer key will help you to ensure that the test is not too easy or too difficult.

Can you write various types of problems using positive and negative numbers?

The Fraction-Decimal Connection

Applying Skills

Write each fraction as a decimal. Then write it as an amount of money using a dollar sign.

1. $\frac{2}{100}$

2. $\frac{48}{100}$

3. $\frac{99}{100}$

4. $\frac{50}{100}$

5. $\frac{2}{10}$

6. $\frac{8}{10}$

Write the value of each group of coins in cents. Write this value as a fraction over 100 pennies. Simplify the fraction. Then, write the value as an amount of money using a dollar sign.

7. 10 pennies

8. 5 nickels

9. 6 dimes

10. 1 quarter, 5 dimes, and 12 pennies

11. 10 dimes, 1 quarter, and 2 nickels

Write the fraction represented by the shaded area of each square or squares. Then, write the corresponding decimal.

12.

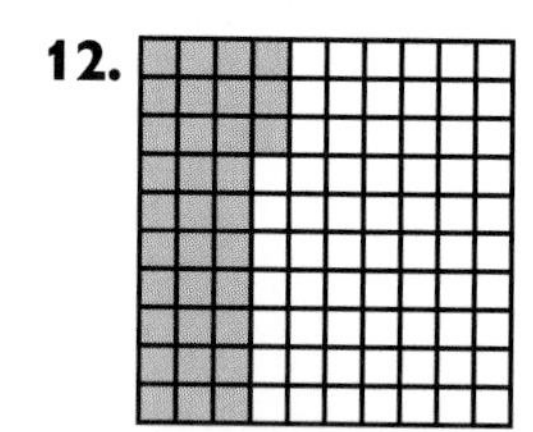

13.

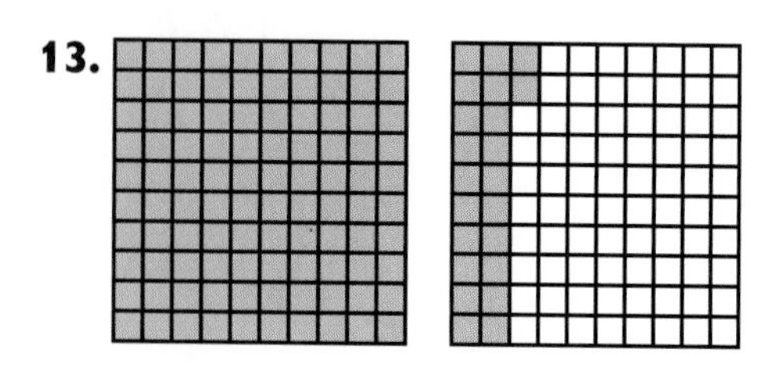

Extending Concepts

Write the fraction represented by the shaded area of each figure. Then, write the corresponding decimal.

14.

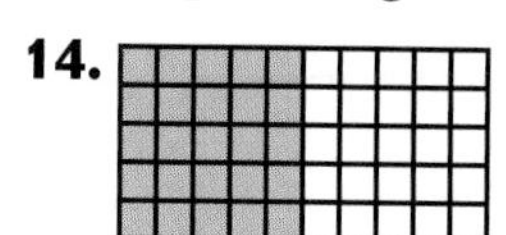

15.

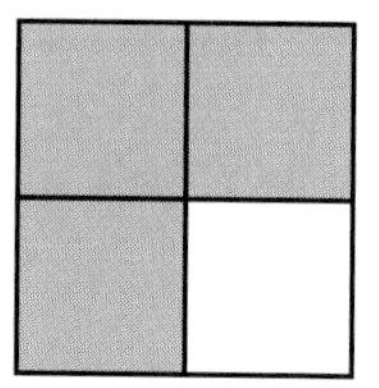

16.

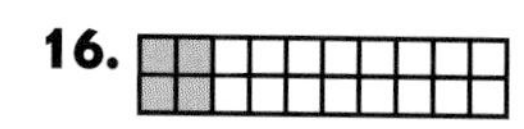

17.

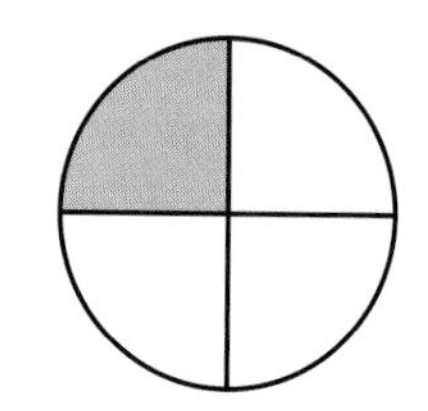

18. This cash register receipt was torn. How much change did the customer get back from the dollar she paid?

Grocery: Energy bar	**$0.88**
Cash	**$1.00**
Change	

Making Connections

A *millisecond* is a thousandth of a second, and a *millicurie* is a thousandth of a curie (a unit of radioactivity). A *mill* is a monetary unit equal to $\frac{1}{10}$ of a cent.

19. What fraction of a dollar is a mill? Give your answer as a fraction and then as a decimal.

20. Explain why you think this unit is called a mill.

What's the Point?

Applying Skills

Write each number.

1. a number with a 3 in the tenths place, a 1 in the hundreds place, a 2 in the tens place, and a 9 in the ones place
2. a number with a 4 in the hundreds place, a 5 in the ones place, a 6 in the tenths place, an 8 in the tens place, and a 3 in the hundredths place
3. a number with a 7 in the thousandths place, a 4 in the ones place, a 1 in the tens place, and zeroes in the tenths and hundredths places
4. a number that is equal to 29.05, but has a digit in its thousandths place

Write each number as a decimal.

5. $5\frac{3}{10}$
6. $137\frac{56}{100}$
7. $10\frac{9}{10}$
8. $6\frac{7}{100}$
9. $12\frac{13}{50}$
10. $4\frac{4}{5}$
11. $5\frac{1}{2}$
12. $12\frac{1}{4}$
13. Complete the pattern by filling in the missing numbers.

$\underline{\ ?\ } \div 4 = 500$ $\qquad \frac{2{,}000}{4} = 500$

$200 \div 4 = 50$ $\qquad \underline{\ ?\ } = 50$

$20 \div 4 = 5$ $\qquad \frac{20}{4} = 5$

$2 \div \underline{\ ?\ } = \underline{\ ?\ }$ $\qquad \frac{2}{4} = \underline{\ ?\ }$

Write each number as a decimal. Use a calculator.

14. $\frac{7}{8}$
15. $\frac{3}{16}$
16. $\frac{25}{200}$
17. $2\frac{3}{4}$
18. $7\frac{4}{5}$
19. $30\frac{23}{40}$
20. $\frac{11}{32}$
21. $\frac{19}{64}$
22. $3\frac{55}{200}$
23. $13\frac{35}{80}$

Extending Concepts

24. Write twelve thousandths as a decimal and as a fraction in simplest form.
25. The egg of the Vervain hummingbird weighs about $\frac{128}{10{,}000}$ ounce. Write this as a decimal. Then, write the decimal in words.

Writing

26. The number 0.52 is read "fifty-two hundredths." The 5 is in the tenths place, and the 2 is in the hundredths place. Explain why these place value positions are named as they are.

Put Them in Order

Applying Skills

Six numbers are graphed on the number line below. Write each number as a fraction, mixed number, or whole number. Then, write each number in decimal notation.

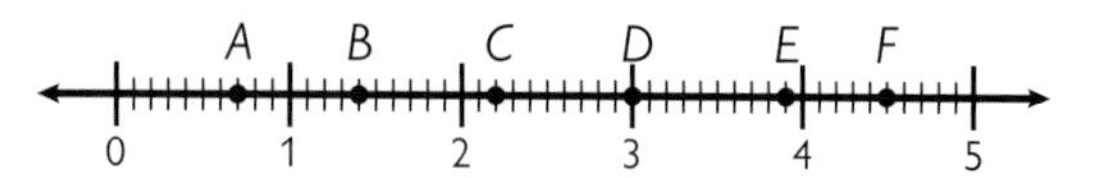

1. point A **2.** point B

3. point C **4.** point D

5. point E **6.** point F

Compare each pair of decimals. Write an expression using <, >, or = to show the comparison.

7. 6.4 and 6.7

8. 5.8 and 12.2

9. 7.02 and 7.20

10. 13.9 and 13.84

11. 16.099 and 160.98

12. 0.331 and 0.303

13. 47.553 and 47.5

Place each set of numbers in order from least to greatest. If two numbers are equal, place an equals sign between them.

14. 14.8, 14.09, 4.99, 14.98, 14.979, 14.099

15. 43, 42.998, 43.16, 42.022, 43.1600, 43.6789

16. 12.3, 12.008, 1.273, 12.54, 120, 12.45

17. 1.2, 4.4. 1.1. 0.9, 17.7, 1.3, 0.95

Extending Concepts

Copy the number line below. Then graph each number.

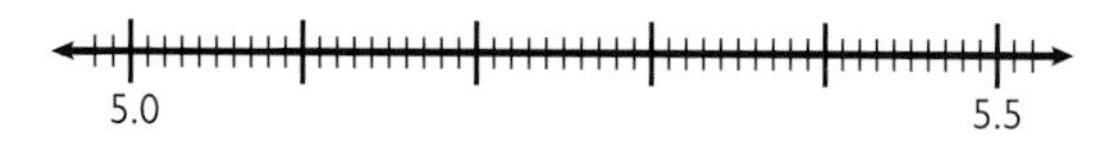

18. 5.07 **19.** 5.24

20. 5.36 **21.** 5.1

22. 5.17 **23.** 5.51

24. Copy the layout below. Place the digits 0, 2, 3, 4, 5, 6, 7, and 8 in the blanks so that the first number is the greatest possible number and the second number is the least possible number. Each digit can be used only once.

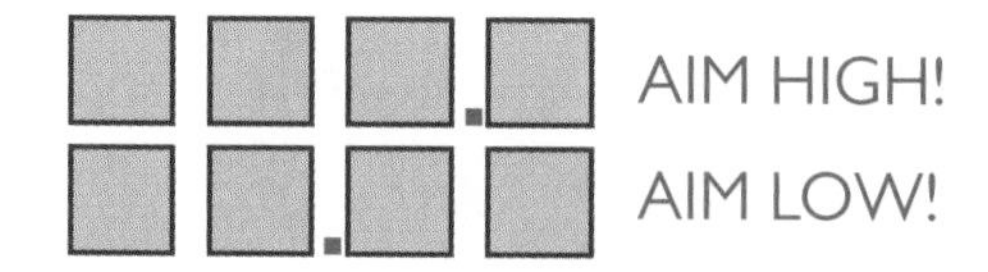

Writing

25. Suppose you are helping a student from another class compare decimals. Write a few sentences explaining why 501.1 is greater than 501.01.

Get It Close Enough

Applying Skills

Copy the number line below. Then graph each number. You may need to estimate some points.

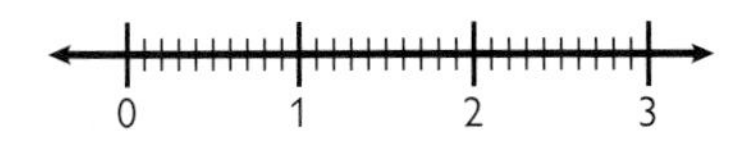

1. 2.5 **2.** 0.75

3. 2.80 **4.** 0.07

5. 0.25 **6.** 1.005

Determine whether each number is between 4.2 and 4.22.

7. 4.06 **8.** 4.217

9. 4.27 **10.** 4.022

11. 4.2016 **12.** 4.2301

13. 4.2099 **14.** 4.199

15. Round 3.2808 to the nearest tenth.

16. Round 3.2808 to the nearest thousandth.

17. Round 33.81497 to the nearest thousandth.

18. Round 33.81497 to the nearest hundredth.

19. Round 4.6745 to the nearest thousandth.

20. Round 4.6745 to the nearest hundredth.

21. Round 4.6745 to the nearest tenth.

22. Round 0.219 to the nearest hundredth.

23. Round 6.97 to the nearest tenth.

24. Round 19.98 to the nearest tenth.

25. Write each of the following numbers on small pieces of paper so you can move them around. Arrange the numbers from least to greatest. If two numbers are equal, place an equals sign between them. Then, place < between the other numbers.

$\frac{6}{10}$	6.10	$6\frac{1}{5}$
0.6666	6.2	$\frac{66}{100}$
$\frac{6}{3}$	$\frac{61}{10}$	0.6667

Extending Concepts

26. Write a number greater than 3.8 that rounds to 3.8. Then, write a number less than 3.8 that rounds to 3.8.

27. Write three numbers that each round to 14.37.

Making Connections

In most countries, people do not measure in inches, feet, and yards. They measure in meters. One meter is a little longer than one yard. $\frac{1}{100}$ of a meter is a centimeter (cm). $\frac{1}{1,000}$ of a meter is a millimeter (mm).

Use a metric ruler to draw a line representing each length.

28. 3.2 cm (32 mm)

29. 8.1 cm (81 mm)

30. 12.6 cm (126 mm)

Place the Point

Applying Skills

Find each sum or difference.

1. 1.25 + 0.68
2. 13.82 − 5.52
3. 15.3 − 0.92
4. 16.89 − 2.35
5. 2.0034 + 25.4
6. 0.007 + 23.6
7. 34.079 − 13.24
8. 16.923 + 2.3
9. 0.89 − 0.256
10. 5 + 2.35
11. 20 − 5.98
12. 17.9 + 7.41

13. Asako and her family are planning a camping trip with a budget of $500.00. They would like to buy the items listed below.

sleeping bag	$108.36
gas stove	$31.78
tent	$359.20
cook set	$39.42
first-aid kit	$21.89
Global Positioning System	$199.99
compass	$26.14
binoculars	$109.76
knife	$19.56
lantern	$20.88

What different combinations of camping equipment could Asako's family afford with the money they have? Name as many combinations as you can. Show your work.

Extending Concepts

14. When Miguel adds numbers in his head, he likes to use expanded notation. For example, 1.32 + 0.276 can be interpreted as 1 + 0 = 1, 0.3 + 0.2 = 0.5, 0.02 + 0.07 = 0.09, and 0.000 + 0.006 = 0.006. The sums are added to give the total of 1.596. Explain what Miguel has done and why it works.

Give the next two numbers for each sequence. Explain how you found the numbers.

15. 2.5, 3.25, 4, 4.75, 5.5, 6.25, . . .

16. 24.8, 23.7, 22.6, 21.5, 20.4, 19.3, . . .

Making Connections

Use the following information from *The World Almanac* about springboard diving at the Olympic Games.

Year	Name/Country	Points
1972	Vladimir Vasin/USSR	594.09
1976	Phil Boggs/U.S.	619.52
1980	Aleksandr Portnov/USSR	905.02
1984	Greg Louganis/U.S.	754.41
1988	Greg Louganis/U.S.	730.80
1992	Mark Lenzi/U.S.	676.53
1996	Xiong Ni/China	701.46
2000	Xiong Ni/China	708.72

17. How many more points did Greg Louganis score in 1984 than in 1988?

18. How many more points did Xiong Ni score in 2000 than in 1996?

More to the Point

Applying Skills

Find each product.

1. 23.62×100
2. 1.876×10
3. $16.8 \times 1{,}000$
4. 78.2×100
5. 125×0.1
6. 56×0.1
7. $7{,}834 \times 0.01$
8. 8×0.01
9. 159×0.001
10. $1{,}008 \times 0.01$

11. Explain how knowing a way to multiply by 0.1, 0.01, and 0.001 can help you multiply by decimals. Use examples if needed.

12. If $670 \times 91 = 60{,}970$, what is the value of 670×9.1?

13. If $1{,}456 \times 645 = 939{,}120$, what is the value of $1{,}456 \times 6.45$?

14. If $57 \times 31 = 1{,}767$, what is the value of 57×0.031?

Find each product.

15. 550×0.3
16. 71×2.2
17. 45×1.1
18. 231×0.12
19. 235×6.2
20. $1{,}025 \times 0.014$
21. 9×25.8
22. 62×0.243
23. $7{,}000 \times 1.8$
24. 41×1.15

25. Choose one of the items 15–24 and explain how you decided where to place the decimal point.

Extending Concepts

26. For Kate's birthday, she and eight of her friends went ice skating. Her mother, father, and grandmother went with them. They took a break for hot chocolate and coffee. The hot chocolate cost $1.35 per cup, and the coffee cost $1.59 per cup. If each of the 9 children had hot chocolate and each of the 3 adults had coffee, how much was the bill?

Give the next two numbers for each sequence. Explain how you found the numbers. You may want to use a calculator.

27. 15.8, 31.6, 63.2, 126.4, 252.8, 505.6, . . .

28. 6.1, 30.5, 152.5, 762.5, 3,812.5, 19,062.5, . . .

29. Write a multiplication sequence of your own. Start with a decimal. Include at least six numbers and the rule for the sequence.

Writing

30. Answer the letter to Dr. Math.

Dear Dr. Math,
What is all the hoopla about the decimal point? I mean, is it really all that important? If it is, why?
Dec-Inez

Decimal Pinpoint

Applying Skills

1. If 382 × 32 = 12,224, what is the value of 38.2 × 0.032?
2. 62 × 876 = 54,312, what is the value of 6.2 × 0.876?
3. If 478 × 52 = 24,856, what is the value of 4.78 × 5.2?
4. If 14 × 75 = 1,050, what is the value of 0.14 × 7.5?

Find each product.

5. 15.2 × 3.4
6. 6.7 × 0.04
7. 587 × 3.2
8. 4.2 × 0.125
9. 0.35 × 1.4
10. 5.2 × 0.065
11. 3.06 × 4.28
12. 0.9 × 0.15
13. 18.37 × 908.44
14. 0.003 × 0.012
15. Choose one of the items 5–14 and explain how you decided where to place the decimal point.
16. Rebecca and Miles are on the decoration committee for the school dance. They need 12 rolls of streamers and 9 bags of balloons. How much will these items cost if a roll of streamers costs $1.39 and a bag of balloons cost $2.09?
17. Pluto's average speed as it travels around the Sun is 10,604 miles per hour. Earth travels 6.28 times faster than Pluto. What is the average speed of Earth?
18. The giant tortoise can travel at a speed of 0.2 kilometer per hour. At this rate, how far can it travel in 1.5 hours?

Extending Concepts

Stacy and Kathryn decided to modify the rules of the "Place Value" game. They used each of the digits 1 to 6 once.

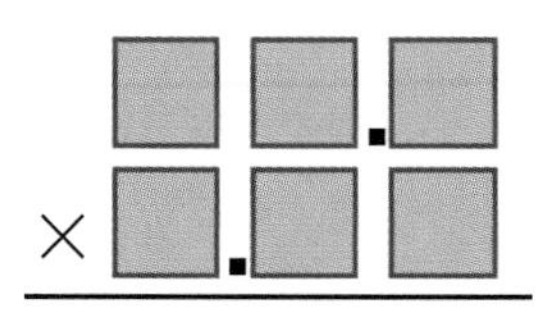

19. How could they place the digits to get the greatest product? What is the greatest product?
20. How could they place the digits to get the least product? What is the least product?

The area of a rectangle equals the length times the width. Find the area of each rectangle.

21.

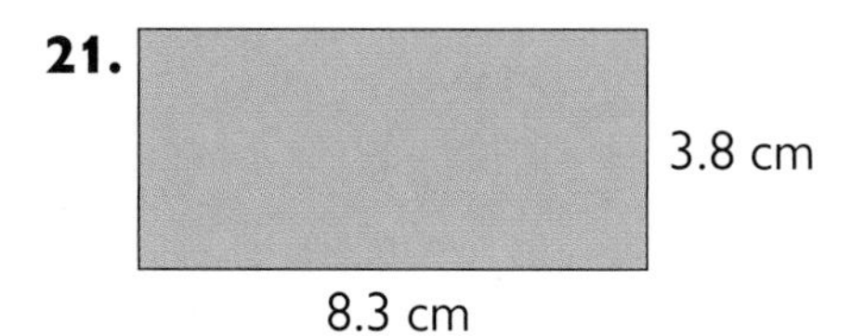

22.

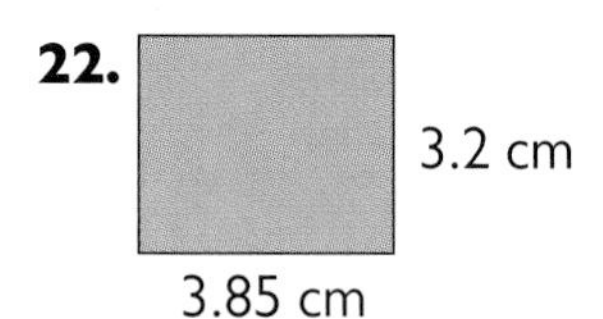

Writing

23. Write some strategies that would help you play the "Place Value" game successfully.

Patterns and Predictions

Applying Skills

1. If 125 ÷ 5 = 25, what is the value of 12.5 ÷ 0.5?

2. If 288 ÷ 12 = 24, what is the value of 2.88 ÷ 0.12?

3. If 369 ÷ 3 = 123, what is the value of 36,900 ÷ 300?

4. If 18,000 ÷ 36 = 500, what is the value of 180 ÷ 0.36?

5. Explain how you can change a decimal division problem to make the problem easier to solve.

Find each quotient.

6. 812 ÷ 0.4

7. 0.34 ÷ 0.2

8. 20.24 ÷ 2.3

9. 180 ÷ 0.36

10. 23 ÷ 0.023

11. 576 ÷ 3.2

12. 14.4 ÷ 0.12

13. 4.416 ÷ 19.2

14. 259.2 ÷ 6.48

15. 4.6848 ÷ 0.366

16. 97.812 ÷ 1.1

17. 38.57 ÷ 1.9

18. 199.68 ÷ 9.6

19. 131.1 ÷ 13.8

20. 5.992 ÷ 74.9

21. 39.95 ÷ 799

Extending Concepts

22. Vladik's dad filled the gasoline tank in his car. The gasoline cost $1.48 per gallon. The total cost of the gasoline was $22.94. How many gallons of gasoline did Vladik's dad put in the tank?

23. A board is 7.5 feet long. If it is cut into pieces that are each 2.5 feet long, how many pieces will there be?

24. Ann and her mom went grocery shopping. Her mom bought 2 dozen oranges for $2.69 per dozen. Estimate the cost of each orange. Explain your reasoning.

25. Mr. and Mrs. Francisco are buying their first home. During the first year, their total mortgage payments will be $12,159.36. How much will their monthly mortgage payment be? Show your work.

26. Drew wants to invite as many friends as he can to go to the movies with him. He has $20.00.

a. The cost of each ticket is $4.50. Estimate how many friends he can take to the movies.

b. A bag of popcorn costs $1.75. Estimate how many friends he can take to the movies if he is going to buy everyone, including himself, a bag of popcorn.

Writing

27. Find the product of 5.5 and 0.12. Then, write two related division problems. Explain how the division problems support the rules for dividing decimals.

It Keeps Going and Going

Applying Skills

Find a decimal representation for each fraction. Then, tell if the decimal is *repeating* or *terminating.*

1. $\frac{3}{8}$ **2.** $\frac{2}{3}$

3. $\frac{5}{6}$ **4.** $\frac{7}{16}$

5. $\frac{7}{10}$ **6.** $\frac{8}{9}$

7. $\frac{5}{11}$ **8.** $\frac{17}{25}$

9. $\frac{2}{11}$ **10.** $\frac{1}{3}$

11. $\frac{3}{4}$ **12.** $\frac{1}{8}$

13. $\frac{5}{9}$ **14.** $\frac{3}{6}$

15. $\frac{4}{15}$ **16.** $\frac{41}{50}$

17. $\frac{12}{15}$ **18.** $\frac{1}{7}$

19. $\frac{9}{11}$ **20.** $\frac{8}{12}$

Change each Celsius temperature to a Fahrenheit temperature.

21. 25°C **22.** 100°C

23. 55°C **24.** 75°C

Change each Fahrenheit temperature to a Celsius temperature.

25. 95°F **26.** 113°F

27. 131°F **28.** 122°F

Extending Concepts

29. One inch is about 2.54 centimeters. Copy and complete the following table.

Name	Height (inches)	Height (centimeters)
John	53	
Krista		127
Miwa		152.4
Tommy	74	

30. Without doing the division, think of three fractions that would have decimal equivalents that repeat. Write each fraction. Use a calculator to confirm that the decimal equivalents are repeating decimals.

Writing

31. Answer the letter to Dr. Math.

Dear Dr. Math,
My partner and I are putting some numbers in order. My partner says that $3.\overline{3}$ is greater than 3.333333. I say that 3.333333 is greater. In fact, it is obvious. Look how many digits it has! Please, tell us who is right and why.
Sincerely,
Endless Lee Confused

Moving to Percents

Applying Skills

For each drawing, determine what percent of the whole is shaded.

1.

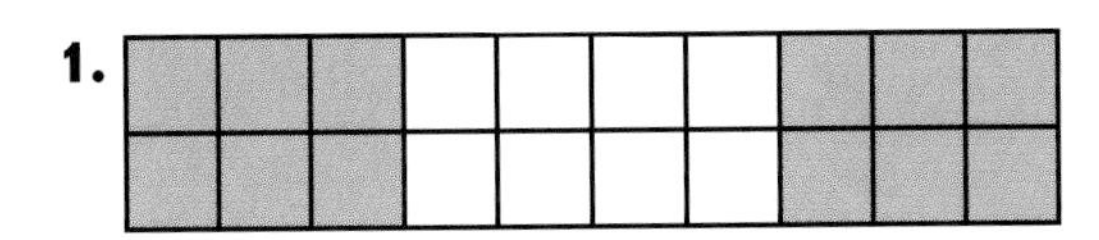

2.

3.

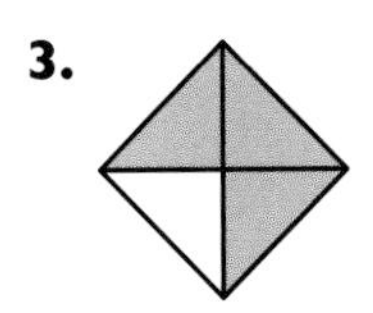

Draw a model showing each percent.

4. 65% **5.** 80% **6.** 30%

7. What percent of the months of the year begin with the letter J?

8. What percent of the months of the year begin with the letter Y?

9. Copy and complete the table.

Fraction	Decimal	Percent
$\frac{4}{10}$		
	0.6	
		90%
		5%
	0.02	
$\frac{1}{8}$		

Extending Concepts

A sixth-grade class is baking cookies for a fund-raiser. They made a graph to show how many of each type were ordered.

10. Write an equivalent fraction for each percent.

11. If a total of 200 cookies were ordered, how many of each type of cookie will the class need to bake?

12. Because macadamia nuts are expensive, the class would lose money for each macadamia nut cookie they sell. Suppose they substitute coconut cookies for orders of macadamia nut cookies. How many coconut cookies will the class need to bake if a total of 200 cookies were ordered?

Writing

13. Answer the letter to Dr. Math.

Dear Dr. Math,
I understand that I can write a fraction as a decimal and a decimal as a fraction. After all, they are both numbers. But, aren't percents different? They have that funny symbol at the end. How can a fraction, a decimal, and a percent all mean the same thing? Could you help me out?
Sincerely,
P.R. Cent

Working with Common Percents

Applying Skills

Match each percent with the closest fraction.

1. 83%	**A.** $\frac{1}{2}$
2. 35%	**B.** $\frac{1}{3}$
3. 27%	**C.** $\frac{2}{3}$
4. 52%	**D.** $\frac{1}{4}$
5. 65%	**E.** $\frac{3}{4}$
6. 42%	**F.** $\frac{1}{5}$
7. 59%	**G.** $\frac{2}{5}$
8. 73%	**H.** $\frac{3}{5}$
9. 19%	**I.** $\frac{4}{5}$

Find 50%, 10%, and 1% of each number.

10. 100	**11.** 600
12. 60	**13.** 40
14. 150	**15.** 340
16. 18	**17.** 44

Extending Concepts

18. Write the numbers $\frac{672}{900}$, 0.012, $\frac{7}{10}$, 32%, $\frac{1}{10}$, 0.721, and 65% in order from least to greatest.

19. Write the numbers 63%, $\frac{2}{10}$, 0.8, $\frac{8}{29}$, 85%, 0.12, $\frac{55}{90}$, and 89% in order from least to greatest.

20. One day a gardener picked about 20% of the strawberries in his garden. Later that day, his wife picked about 25% of the remaining strawberries. Still later, their son picked about 33% of the strawberries that were left. Even later, their daughter picked about 50% of the remaining strawberries in the garden. Finally, there were only 3 strawberries left. About how many strawberries were originally in the garden? Explain your reasoning.

Writing

21. Explain the connection between fractions, decimals, and percents. Show examples.

Percent Power

Applying Skills

Estimate each value.

1. 19% of 30
2. 27% of 64
3. 48% of 72
4. 73% of 20
5. 67% of 93
6. 25% of 41
7. 65% of 76
8. 81% of 31
9. 34% of 301
10. 41% of 39

Find each value.

11. 25% of 66
12. 13% of 80
13. 52% of 90
14. 16% of 130
15. 37% of 900
16. 32% of 68
17. 66% of 43
18. 7% of 92
19. 42% of 85
20. 78% of 125

Extending Concepts

21. Find the total cost of a $25.50 meal after a tip of 15% has been added.

22. A shirt that sold for $49.99 has been marked down 30%. Find the sale price.

23. Twenty-seven percent of Camille's annual income goes to state and federal taxes. If she earned $35,672, how much did she actually keep?

24. Jared's uncle bought 100 shares of stock for $37.25 per share. In six months, the stock went up 30%. How much were the 100 shares of stock worth in six months?

25. Kyal's family bought a treadmill on sale. It was 25% off the original price of $1,399.95. They paid half of the sale price as a down payment. What was the down payment?

26. Maxine and her family went on a vacation to San Francisco last summer. On the first day, they gave the taxi driver a 15% tip and the hotel bellhop a $2.00 tip for taking the luggage to the room. The taxi fare was $15.75. How much did they spend in all?

Making Connections

For a healthy diet, the total calories a person consumes should be no more than 30% fat. Tell whether each food meets these recommendations. Show your work.

27. Reduced-fat chips
Calories per serving: 140
Calories from fat: 70

28. Low-fat snack bar
Calories per serving: 150
Calories from fat: 25

29. Light popcorn
Calories per serving: 30
Calories from fat: 6

30. Healthy soup
Calories per serving: 110
Calories from fat: 25

Homework 13

Less Common Percents

Applying Skills

Write an equivalent decimal and an equivalent fraction for each percent.

1. 0.4%

2. 0.3%

3. 0.25%

4. 0.05%

5. 0.079%

6. 0.008%

7. $\frac{1}{5}\%$

8. $\frac{7}{10}\%$

9. $\frac{1}{4}\%$

10. $\frac{4}{25}\%$

11. $\frac{1}{8}\%$

12. $\frac{1}{25}\%$

State whether each number represents a percent less than one. Write *yes* or *no*.

13. 0.02

14. 0.0019

15. 0.0101

16. 0.7

17. 0.0009

18. 0.0088

19. $\frac{1}{50}$

20. $\frac{1}{2,500}$

21. $\frac{7}{300}$

22. $\frac{2}{300}$

23. $\frac{9}{1,000}$

24. $\frac{4}{700}$

25. Make a model for 110%.

26. Write an equivalent decimal and an equivalent mixed number for 925%.

27. Order 25%, $\frac{1}{4}\%$, 125%, and 1 from least to greatest.

Extending Concepts

28. Liana makes beaded jewelry. She plans to sell them at the school fair. Liana wants to sell each item for 135% of the cost to make the item. Copy and complete the table below to find the selling price of each item.

Item	Cost to Make	Selling Price
Long Necklace	$22.00	
Short Necklace	$16.00	
Bracelet	$14.50	
Earrings	$9.25	

29. During a recent contest at South Middle School, prizes were awarded to 0.4% of the students. If there are 500 students at the school, how many students won prizes?

30. The diameter of the Sun is 865,500 miles. The diameter of Earth is about 0.9% of the Sun's diameter. Find the diameter of Earth.

Writing

31. A coach wants his players to give 110% of their effort. Explain what is meant by 110%. Is it reasonable for a coach to ask for 110% of their effort? Explain.

Give It to Me Straight

Applying Skills

Examine each statement. Determine whether or not the percent is used correctly. Explain your answer.

1. Unemployment fell 5% in July.
2. Yogi Berra said, "Half the game is 90% mental."
3. Roberto gave away 130% of his stamp collection.
4. 72% of the class prefers snacks that contain chocolate.
5. The team won 70% of the games, lost 25% of the games, and tied 10% of the games.
6. Your height now is 0.1% of your height at one year old.
7. The store's sales for the month of April are 130% of its sales for the month of March.
8. The value of the house is 125% its value 5 years ago.
9. 45% of the class members have a brother, 50% of the class members have a sister, and 20% of the class members have no brother or sister.
10. At the class picnic, the students ate 130% of the cookies prepared by the parents.

Extending Concepts

11. Mark and Fala collected data at lunchtime to determine which types of desserts were purchased most often. The results of the 75 students who went through the lunch line are given below.

Type of Dessert	Percent of Students who Purchased the Dessert
Cookies	24%
Ice Cream	40%
Fruit	4%
Candy	12%
None	20%

How many students bought each type of dessert? How many students bought no dessert?

12. North Middle School has 645 students equally divided into grades six, seven, and eight. During a recent school survey, Dan noticed that only eighth graders played lacrosse. After analyzing the data, he said, "86 students play lacrosse." Paula said, "No way! 40% of the eighth graders play lacrosse." Explain how both students can be correct.

Writing

13. Throughout this unit, you have used fractions, decimals, and percents. Explain how the three are connected. Describe the advantage of each representation. Use examples to help to clarify your explanation.

The Other End of the Number Line

Applying Skills

Determine whether each sentence is *true* or *false.*

1. $4 > -2$

2. $3\frac{1}{2} < 3\frac{3}{4}$

3. $0.3 > 0.25$

4. $0 > -1$

5. $-\frac{1}{4} < -\frac{1}{2}$

6. $-5 > -4$

7. $-6.5 > -6$

8. $-107 < -106$

Use $<$ or $>$ to compare each pair of numbers.

9. 37 and 42

10. -25 and -37

11. -12 and 12

12. -144 and -225

13. -512 and -550

14. -300 and -305

15. -960 and -890

16. 385 and 421

Evan and Alison are playing "Who's Got My Number?" They are reading cards to each other. Determine the next number for each card.

17. My number is -13. Does anyone have 4 more?

18. My number is -21. Does anyone have 25 more?

19. My number is -9. Does anyone have 9 more?

20. My number is -14. Does anyone have 7 less?

21. My number is 4. Does anyone have 17 less?

22. Arrange the answers from items 17–21 in order from least to greatest.

Extending Concepts

Order each set of numbers from least to greatest.

23. $2, -2.5, -6.8, 6.7, 4, 0.731, -3$

24. $-\frac{1}{3}, 4, -\frac{3}{4}, \frac{7}{8}, 1, -\frac{7}{8}, -2, 5$

25. $-1\frac{2}{3}, -\frac{6}{4}, 3, 0, 4\frac{1}{5}, \frac{5}{3}, 4, -2, -1\frac{4}{5}$

Making Connections

26. Consider the elevations listed below. List the elevations from greatest to least.

Location	Elevation (meters)
Dead Sea, Israel	−408
Chimborazo, Ecuador	6,267
Death Valley, California	−86
Challenger Deep, Pacific Ocean	−10,924
Zuidplaspoldor, Netherlands	−7
Fujiyama, Japan	3,776

27. Find the elevations of three other interesting places around the world. Add these elevations to the list in item **26**.

Homework 16

Moving on the Number Line

Applying Skills

Write an addition equation for each number line.

1.

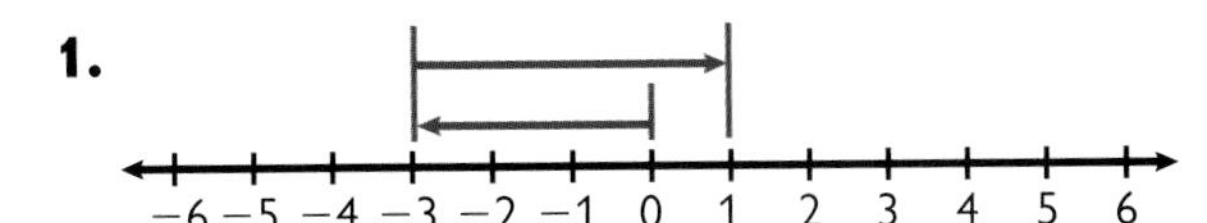

2.

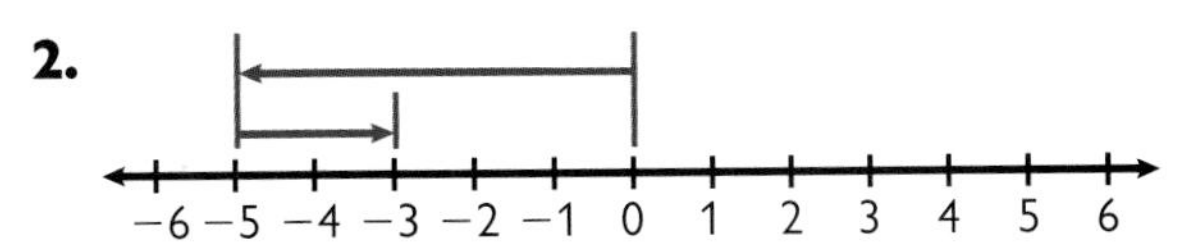

3.

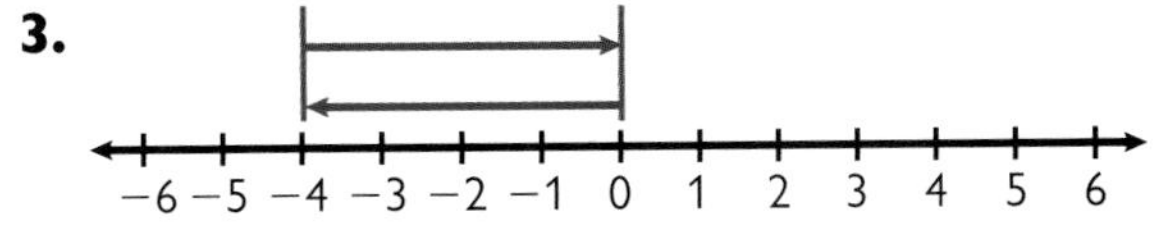

4.

−6 −5 −4 −3 −2 −1 0 1 2 3 4 5 6

5.

−6 −5 −4 −3 −2 −1 0 1 2 3 4 5 6

Use a number line to find each sum.

6. $-4 + 7$ **7.** $6 + (-3)$

8. $-3 + (-2)$ **9.** $-2 + (-2)$

10. $1 + (-5)$ **11.** $-8 + 5$

12. $-5 + 5$ **13.** $-5 + 7$

14. $6 + (-9)$ **15.** $-2 + (-5)$

Extending Concepts

Fractions and decimals also have negatives. Use what you have learned to find each sum.

16. $1\frac{3}{5} + (-5)$ **17.** $-\frac{3}{7} + (-\frac{6}{7})$

18. $5.2 + (-7.3)$ **19.** $-12.352 + 4.327$

Writing

20. Answer the letter to Dr. Math.

Dear Dr. Math,

The other day in math class, I was finding -3 + 7 + (-5) + 1. I started drawing arrows on the number line. Then, my friend Maya walked by and said the answer was zero. I asked how she got the answer so fast. She told me that she added the two positive numbers (7 + 1) and got 8. Then, she added the two negative numbers (-3 + (-5)) and got -8. She said 8 + (-8) is zero. I thought she was wrong, but when I checked it on the number line, the answer was zero. Why is it OK to add the numbers out of order?

Signed,

Addled About Addition

Homework 17 Taking the Challenge

Applying Skills

Margo (pink) and Lisa (green) are playing "Color Challenge." For each set of spins, determine who is ahead. Then determine how much the person is ahead.

1. Add 2 green cubes.
Add 1 pink cube.
Add 1 green cube.
Add 2 pink cubes.

2. Add 1 green cube.
Add 1 green cube.
Add 3 pink cubes.
Add 2 green cubes.
Add 1 pink cube.
Add 2 pink cubes.

3. Add 2 pink cubes.
Add 1 pink cube.
Add 1 pink cube.
Add 2 green cubes.
Add 1 green cube.
Add 2 pink cubes.
Add 3 green cubes.
Add 1 green cube.

Write an addition equation for each drawing.

4.

5.

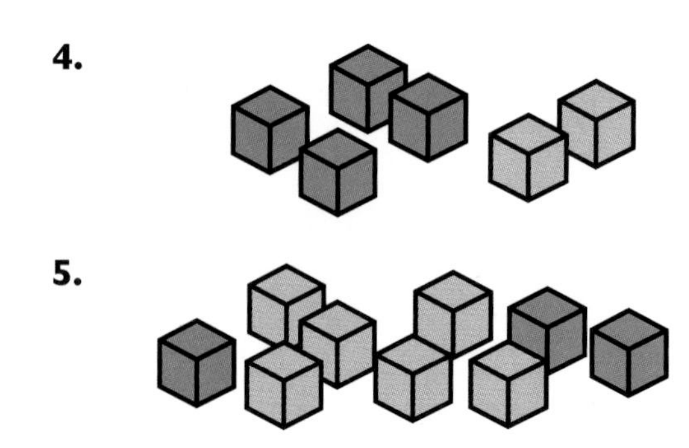

Use cubes to find each sum.

6. $-2 + 6$ **7.** $-5 + 3$

8. $4 + (-5)$ **9.** $7 + (-4)$

10. $-5 + (-6)$ **11.** $-1 + (-8)$

Extending Concepts

You may know that multiplying is just a fast way of adding. For example, you can use 4×3 to find $3 + 3 + 3 + 3$.

12. Consider $-3 + (-3) + (-3) + (-3)$.

a. Draw a collection of cubes that represent the addition problem.

b. Write a corresponding multiplication problem and its product.

13. Consider $-2 + (-2) + (-2) + (-2) + (-2)$.

a. Draw a collection of cubes that represent the addition problem.

b. Write a corresponding multiplication problem and its product.

14. What is $4 \times (-5)$? Try to find the answer without using drawings or cubes.

Making Connections

15. At the end of last month, Keston had $1,200 in his savings account. His next bank statement listed a deposit of $100, a withdrawal of $300, and another deposit of $150. Write an addition problem to represent the situation. How much money does he have in his account now?

The Meaning of the Sign

Applying Skills

Write a subtraction equation for each number line.

1.

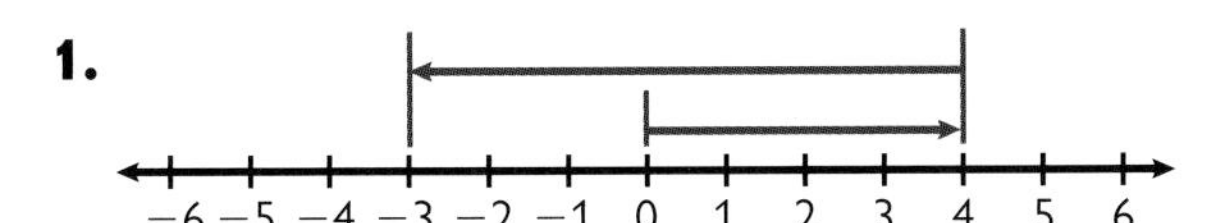

2.

−6 −5 −4 −3 −2 −1 0 1 2 3 4 5 6

3.

−6 −5 −4 −3 −2 −1 0 1 2 3 4 5 6

4.

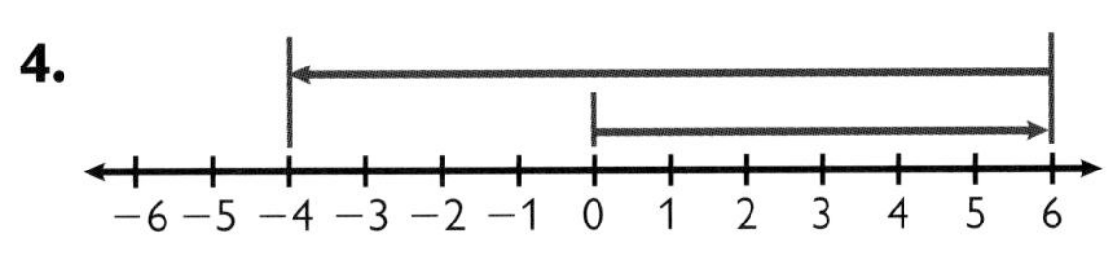

5.

−6 −5 −4 −3 −2 −1 0 1 2 3 4 5 6

Use a number line to find each difference.

6. $3 - 8$

7. $-1 - 5$

8. $-3 - 2$

9. $-2 - 2$

10. $5 - 9$

11. $-2 - 9$

12. $-5 - 4$

13. $6 - 8$

14. $-7 - 8$

15. $4 - 10$

Extending Concepts

You must add and subtract integers according to the order of operations. For each problem, two students got different answers. Decide which student has the correct answer.

16. Ellen: $-3 - 7 + 1 = -11$
Tia: $-3 - 7 + 1 = -9$

17. Robert: $4 - 3 + 6 = 7$
Josh: $4 - 3 + 6 = -5$

18. Emma: $-2 + (-4) - 7 + (-1) = -12$
Lucas: $-2 + (-4) - 7 + (-1) = -14$

19. HaJeong: $3 - 5 + (-3) - 1 = 0$
Janaé: $3 - 5 + (-3) - 1 = -6$

Making Connections

20. Opal and Dory are piloting a deep sea submersible at the depth of 200 meters below sea level. To retrieve some dropped equipment, they must descend another 2,400 meters. Write a subtraction equation that describes the situation. What is the final depth?

21. The afternoon temperature was 15°F. During the night, the temperature dropped 17°F. Write a subtraction equation that describes the situation. What was the nighttime temperature?

The Cube Model

Applying Skills

Write a subtraction equation for each drawing.

1.

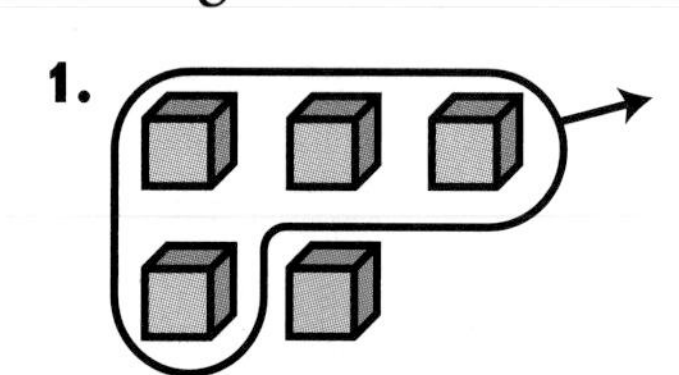

2.

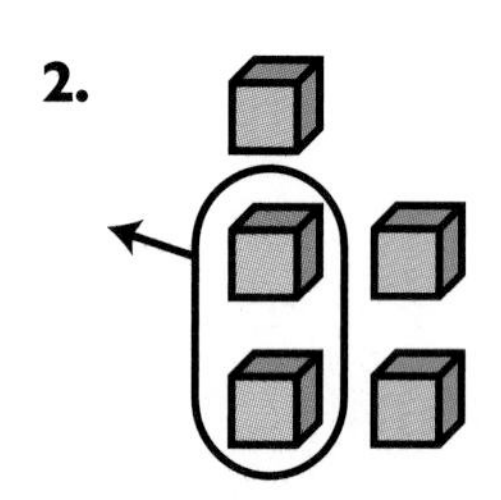

3.

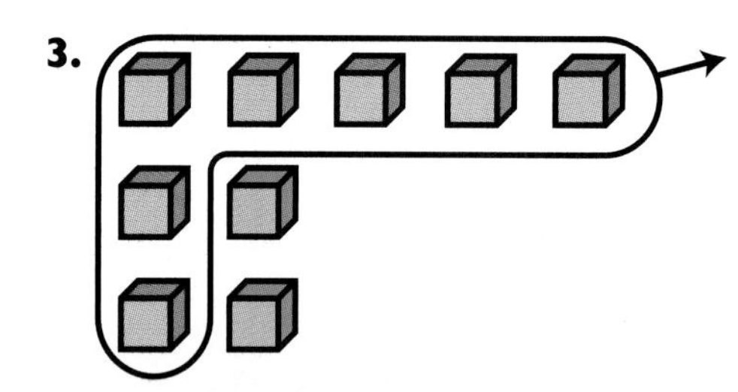

4.

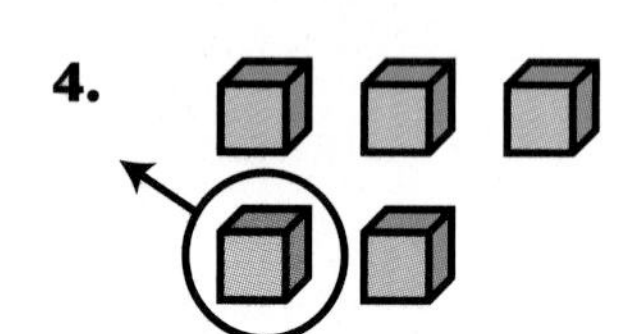

5.

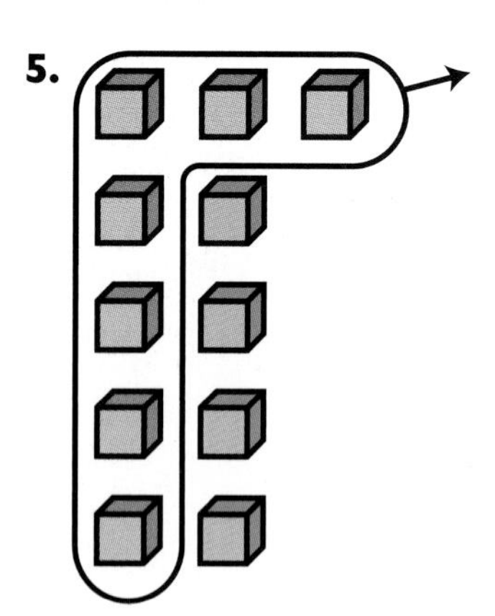

Use cubes to find each difference.

6. $-2 - 6$
7. $5 - 7$
8. $-5 - 2$
9. $4 - 7$
10. $-1 - 1$
11. $-1 - 3$
12. $6 - 9$
13. $-5 - 6$
14. $-8 - 2$
15. $8 - 9$

Extending Concepts

You may have thought of the division problem $15 \div 5$ as dividing 15 things into 5 groups.

16. Suppose you have 12 negative cubes.
    a. Make a drawing to show how the cubes can be divided into 3 groups.
    b. Write a division problem for the situation.
    c. What is the quotient?
17. Suppose you have 10 negative cubes.
    a. Make a drawing to show how the cubes can be divided into 5 groups.
    b. Write a division problem for the situation.
    c. What is the quotient?
18. What is $-24 \div 6$? Try to find the answer without using drawings or cubes. Explain your answer.

Writing

19. Use words and/or pictures to describe how you would use zero pairs to find $-4 - 6$.

Write It another Way

Applying Skills

Write each subtraction problem as an addition problem. Then, solve the problem.

1. $-5 - 1$ **2.** $4 - 11$

3. $-2 - 7$ **4.** $8 - 6$

5. $-1 - 8$ **6.** $0 - 7$

Write each addition problem as a subtraction problem. Then, solve the problem.

7. $-4 + (-5)$ **8.** $8 + (-7)$

9. $3 + (-11)$ **10.** $-7 + (-8)$

11. $-1 + (-5)$ **12.** $6 + (-6)$

Find each sum or difference.

13. $-3 + 8$ **14.** $-3 - 8$

15. $-4 + (-6)$ **16.** $5 - 9$

17. $0 - 2$ **18.** $-3 + 10$

Extending Concepts

19. Use the subtraction pattern you started at the beginning of the lesson.

a. Extend the list of subtraction problems to include the following:

$4 - (-1)$

$4 - (-2)$

$4 - (-3)$

$4 - (-4)$

$4 - (-5)$

b. Use the pattern to find the solutions to the subtraction problems.

c. Copy and complete the following sentence.

Subtracting a negative number gives the same answer as if you __?__.

Use what you have learned from the pattern to find each difference.

20. $5 - (-3)$ **21.** $-3 - (-7)$

22. $-8 - (-4)$ **23.** $1 - (-8)$

24. $0 - (-3)$ **25.** $-1 - (-6)$

26. $9 - (-5)$ **27.** $7 - (-7)$

Making Connections

28. Death Valley, California, is the lowest elevation in the United States. It is 280 feet below sea level, or -280 feet in elevation. Mt. McKinley, Alaska, has the highest elevation at about 20,000 feet above sea level. Write a number sentence that shows the difference in elevation between Mt. McKinley and Death Valley. What is the difference?

29. The highest temperature recorded in North America is 134°F in Death Valley, California. The lowest recorded temperature is -87°F in Northice, Greenland. What is the difference in temperatures?

The McGraw·Hill Companies

This unit of MathScape: Seeing and Thinking Mathematically was developed by the Seeing and Thinking Mathematically project (STM), based at Education Development Center, Inc. (EDC), a non-profit educational research and development organization in Newton, MA. The STM project was supported, in part, by the National Science Foundation Grant No. 9054677. Opinions expressed are those of the authors and not necessarily those of the Foundation.

CREDITS: **206** (tl)Glencoe file photo, (tr)Getty Images. **207** (tl tc tr)Getty Images, (b)Najlah Feanny/CORBIS; **208 218** Getty Images; **229** Matt Meadows; **230** Getty Images; **237** David Young-Wolff/PhotoEdit; **239** Chris Windsor/Getty Images; **241 242** Getty Images; **249** Aaron Haupt.

Send all inquiries to:
Glencoe/McGraw-Hill
8787 Orion Place
Columbus, OH 43240-4027

ISBN: 0-07-866800-X

4 5 6 7 8 9 10 058 06